About Island Press

Since 1984, the nonprofit organization Island Press has been stimulating, shaping, and communicating ideas that are essential for solving environmental problems worldwide. With more than 800 titles in print and some 40 new releases each year, we are the nation's leading publisher on environmental issues. We identify innovative thinkers and emerging trends in the environmental field. We work with world-renowned experts and authors to develop cross-disciplinary solutions to environmental challenges.

Island Press designs and executes educational campaigns in conjunction with our authors to communicate their critical messages in print, in person, and online using the latest technologies, innovative programs, and the media. Our goal is to reach targeted audiences—scientists, policymakers, environmental advocates, urban planners, the media, and concerned citizens—with information that can be used to create the framework for long-term ecological health and human well-being.

Island Press gratefully acknowledges major support of our work by The Agua Fund, The Andrew W. Mellon Foundation, Betsy & Jesse Fink Foundation, The Bobolink Foundation, The Curtis and Edith Munson Foundation, Forrest C. and Frances H. Lattner Foundation, G.O. Forward Fund of the Saint Paul Foundation, Gordon and Betty Moore Foundation, The Kresge Foundation, The Margaret A. Cargill Foundation, New Mexico Water Initiative, a project of Hanuman Foundation, The Overbrook Foundation, The S.D. Bechtel, Jr. Foundation, The Summit Charitable Foundation, Inc., V. Kann Rasmussen Foundation, The Wallace Alexander Gerbode Foundation, and other generous supporters.

The opinions expressed in this book are those of the author(s) and do not necessarily reflect the views of our supporters.

Our Renewable Future

Our Renewable Future

Laying the Path for 100% Clean Energy

By Richard Heinberg and David Fridley

Post Carbon Institute
Santa Rosa, California

ISLANDPRESS

Washington | Covelo | London

ISLAND PRESS is a trademark of the Center for Resource Economics.

Library of Congress Control Number: 2016931452

ISBN-13: 978-1-61091-779-7 (paper)
ISBN-13: 978-1-61091-780-3 (electronic)

Printed on recycled, acid-free paper

Manufactured in the United States of America
10 9 8 7 6 5 4 3

Keywords: Biomass, cap and trade, carbon capture, carbon tax, coal, consumerism, electric cars, electricity, energy efficiency, energy grid, energy storage, energy supply, fossil fuels, geothermal, gross domestic product, hydrogen, hydropower, industrialization, natural gas, net energy, nuclear, oil, petroleum, photovoltaic cells, renewable energy, solar, solar panels, wind, wind turbines

Contents

Figures and Table

Figures

Table

Acknowledgments

The authors extend their deepest thanks to Post Carbon Institute staff for their work on this book: Daniel Lerch for his extensive editorial efforts; Asher Miller for guiding the project from inception to completion; and Leslie Moyer for coordinating input from contributors. Many thanks also to Rebecca Bright and Sharis Simonian from Island Press for helping make this book a reality; to Adrian Wiegman and Jeff Rutherford for their research assistance; to the participants in our advisor summits held in New York and San Francisco in summer 2015; to Levana Saxon for facilitating those gatherings; and to our many reviewers and contributors, including Philip Ackerman-Leist, David Blittersdorf, Mike Bomford, Mik Carbajales-Dale, Heather Cooley, Casey Haskell, Elena Krieger, Susan Krumdieck, David Morris, Diane Moss, Mateo Nube, Cindy Parker, Anthony Perl, David Pomerantz, James Rose, Brian Schwartz, Bill Sheehan, Stiv Wilson, and Miya Yoshitani. Finally, special thanks to Irene Krarup and the trustees of the V. Kann Rasmussen Foundation for their support of this project, and to PJ and LH for their steadfast support of Post Carbon Institute.

Introduction

THE NEXT FEW DECADES will see a profound and all-encompassing energy transformation throughout the world. Whereas society now derives the great majority of its energy from fossil fuels, by the end of the century we will depend primarily on renewable sources like solar, wind, biomass, and geothermal power.

Two irresistible forces will drive this historic transition.

The first is the necessity of avoiding catastrophic climate change. In December 2015, 196 nations unanimously agreed to limit global warming to no more than two degrees Celsius above preindustrial temperatures.[1] While some of this reduction could technically be achieved by carbon capture and storage from coal power plants, carbon sequestration in soils and forests, and other "negative emissions" technologies and efforts, the great majority of it will require dramatic cuts in fossil fuel consumption.

The second force driving a post-carbon energy shift is the ongoing depletion of the world's oil, coal, and natural gas resources. Even if we do nothing to avoid climate change, our current energy regime remains unsustainable.

Though Earth's crust still holds enormous quantities of fossil fuels, economically useful portions of this resource base are much smaller, and the fossil fuel industry has typically targeted the highest-quality, easiest-to-access resources first.

All fossil fuel producers face the problem of declining resource quality, but the problem is most apparent in the petroleum sector. During the decade from 2005 to 2015, the oil industry's costs of production rose by over 10 percent per year because the world's cheap, conventional oil reserves—the "low-hanging fruit"—are now dwindling (fig. I.1). While new extraction technologies make lower-quality resources accessible (like tar sands, and tight oil from fracking), these technologies require higher levels of investment and usually entail heightened environmental risks. World coal and gas supplies have yet to reach the same higher-cost tipping point; however, several recent studies suggest that the end of affordable supplies of these fuels may be years—not decades—away.[2] We will be consuming fossil fuels for many years to come, no doubt; but their decline is inevitable. We are headed to a nonfossil future whether we're ready or not.

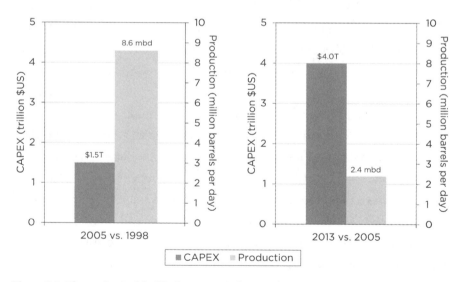

Figure I.1. Change in world oil industry capital expenditures (CAPEX) and crude oil production, 2005 vs. 1998 and 2013 vs. 2005.
Source: U.S. Energy Information Administration and Steven Kopits, "Oil and Economic Growth: A Supply Constrained View," presentation at Center on Global Energy Policy, Columbia University, New York, February 11, 2014.

Nuclear fission power is not likely to play a larger role in our energy future than it does today, outside of China and a few other nations, if current trends continue. Indeed, high investment and (post-Fukushima) safety requirements, growing challenges of waste storage and disposal, and the risks of catastrophic accidents and weapons proliferation may together result in a significant overall shrinkage of the nuclear industry by the end of the century. Despite recent press reports about progress in hot fusion power and claims for "cold fusion," these energy sources currently produce no commercial energy and—even if claims turn out to be justified—they are unlikely to do so on a significant scale for decades to come.

Fossil fuels are on their way out one way or another, and nuclear energy is a dead end. That leaves renewable energy sources, such as solar, wind, hydro, geothermal, and biomass, to shoulder the burden of powering future society. While it is probably an oversimplification to say that people in the not-too-distant future will inhabit a 100 percent renewably powered world, it is worth exploring what a complete, or nearly complete, shift in our energy systems would actually mean. Because energy is implicit not only in everything we do but also in the built environment around us (which requires energy for its construction, maintenance, and disposal/decommissioning), it is in effect the wellspring of our existence. As the world embarks on a transformative change in its energy sources, the eventual impacts may include a profound alteration of people's personal and collective habits and expectations, as well as a transformation of the structures and infrastructure around us. Our lives, communities, and economies changed radically with the transition from wood and muscle power to fossil fuels, and so it is logical that a transition from fossil fuels to renewables— that is, a fundamental change in the quantity and quality of energy available to power human civilization—will also entail a major shift in how we live.

How would a 100 percent renewable world look and feel? How might the great-grandchildren of today's college students move through a typical day without using fossil fuels either directly or indirectly? Where will their food come from? How will they get from place to place? What will the buildings they inhabit look like, and how will those buildings function? Visions of the future are always wrong in detail, and often even in broad strokes; but sometimes they can be wrong in useful ways. Scenario exercises can help us evaluate and prepare for a variety of outcomes, even if we don't know precisely which reality will emerge. Further, by imagining the future we often help create it: advertisers and industrialists long ago learned that creative product developers, marketers, and commercial artists can shape the choices, actions, and expectations of entire societies. If we are embarking upon what may turn out to be history's most significant energy transition, we should spend some effort now to imagine an all-renewable world, even though the exercise will inevitably involve guesswork and oversimplification.

A good way to begin visualizing the post-carbon future might be to explore how and why we came to construct our current "normal" reality of energy consumption.

How "Normal" Came to Be

For most people living in the early nineteenth century, firewood was the dominant fuel and muscles were the primary source of power. The entire economy—including the design of towns and homes, and people's daily routines—was structured to take advantage of the capabilities of wood and human or animal muscle. Food staples were often grown close to the point of consumption in order to minimize the need for slow and expensive horse- or sail-drawn transport. Many people worked as farmers or farm laborers, because many hands were required to do the fieldwork needed to produce sufficient food for the entire population. Traction animals were significant symbols of wealth: a prosperous farmer might own a team of oxen or mules, while his well-off cousin in the city might keep a horse or two to provide personal mobility. In slave-holding portions of the United States, some humans claimed ownership of other humans so as to make economic use of their intelligently directed muscles—a horrific practice that shattered the lives of millions (its effects continue to reverberate) and was ended only by an epic war. Meanwhile, vast tracts of forest in the northeastern United States were being cleared to provide fuel for home heating and, increasingly, for the operation of industrial machinery, including steamboats and steam locomotives.

Agrarian life in the nineteenth century. (Credit: Carl Conrad Dahlberg, Malmö Art Museum, via Wikimedia Commons.)

Then, in the mid-1800s, along came fossil fuels. Compared to firewood, coal and oil were more energy dense and therefore more portable, and they could be made available in greater quantities (especially since forests were disappearing due to overcutting). Compared to muscles, fuel-fed machines were formidable and tireless. Nineteenth-century inventors had already been devising ways to reduce labor through mechanization and to create new opportunities for mobility, communication, and amenity with devices ranging from the telegraph to the rail locomotive. The advent of cheap, abundant, and transportable fossil energy sources encouraged a flood of new or improved energy-consuming technologies.

A series of significant inventions—including the electricity generator, alternating current, and the electric motor—made energy from coal (also from moving water and later from natural gas and nuclear fission) available in homes and offices. This opened the potential for electric lighting, washing machines, vacuum cleaners, and an ever-expanding array of entertainment and communications devices, including telephones, radios, televisions, and computers.

Meanwhile, liquid fuels made from petroleum mobilized the economy as never before. Automobiles, airplanes, trucks, ships, and diesel-fueled trains began hauling people and freight at distances and speeds—and in quantities— that were previously unimaginable. Oil products also began fueling society's raw materials extraction processes—mining, forestry, and fishing—resulting in far higher rates of production at much lower costs. By the mid-twentieth century, oil was increasingly transformed into plastics, chemicals, lubricants, and pharmaceuticals. And oil-powered machinery replaced human labor in agriculture, resulting in one of the most significant demographic shifts in history as the bulk of humanity left farms and moved to cities (fig. I.2).

Because fossil fuels were so cheap relative to the power of muscles, machines took over much of the drudgery of life. Whereas human slavery had figured prominently in parts of the U.S. economy in the early nineteenth century, today each American commands the services of hundreds of "energy slaves"[3] counted as the number of persons whose full-time labor would be required to substitute for the services currently provided by powered machinery.

As energy is consumed in the making of roads, buildings, pipelines, food, clothing, and other products, it is effectively embedded or embodied in those objects. The built environment around us, and the manufactured goods with

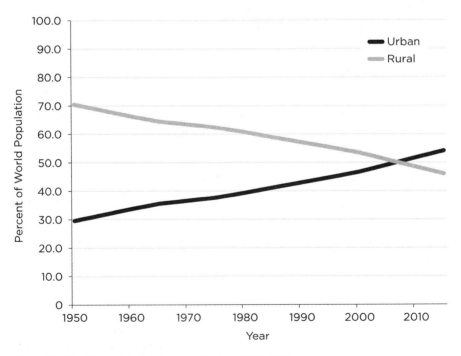

Figure I.2. World rural and urban population, 1950–2015.
Source: United Nations Department of Economic and Social Affairs, "World Urbanization Prospects 2014," http://esa.un.org/unpd/wup/.

which we surround ourselves, represent immense amounts of fossil energy—
energy used in the production of materials and goods through the operations
of mining equipment, smelters, cement makers, trucks, road surfacers, and
factories.

During the same period in which fossil fuels began to power most aspects of
daily life, we began to design our homes and cities to fit the machines and products that used those fuels or embodied the energy from their combustion. The
automobile became the design centerpiece for suburbs, shopping malls, park-
ing lots, garages, and highways. Meanwhile, expansion of transport by airplane
required the construction of airports—the largest of which cover as much space
as was formerly occupied by entire towns.

All of this was undertaken with the tacit assumption that society would always have more fossil energy with which to maintain and operate its ever-expanding infrastructure. There was no long-range grand plan guiding the project. The fossil-fueling of the economy happened bit by bit, each new element
building on the last, with opportunity leading to innovation. What was technically possible became economically necessary . . . and hence normal.

It is easy now to take it all for granted. But we shouldn't. As the energy
sources that built the twentieth century ebb, it may be helpful to disabuse ourselves of many of our assumptions and expectations by observing how different
"normal" is for North Americans as compared with people in rural villages in
less industrialized countries, or by reading first-person narratives of daily life in
the eighteenth and nineteenth centuries. As profoundly dissimilar as our current "normal" is to human experience prior to the industrial revolution, the
future may be just as different again.

Why a Renewable World Will Be Different

Solar, wind, hydro, and geothermal generators produce electricity, and we already have an abundance of technologies that rely on electricity. So why should
we need to change the ways we use energy? Presumably all that's necessary is to
unplug coal power plants, plug in solar panels and wind turbines, and continue
living as we do currently.

This is a misleading way of imagining the energy transition for six important reasons.

1. *Intermittency.* As we will see in chapter 3, the on-demand way we use electricity now is unsuited to variable renewable supplies from solar and wind. Power engineers designed our current electricity production, distribution, and consumption systems around controllable inputs (hydro, coal, natural gas, and nuclear), but solar and wind are inherently uncontrollable: we cannot force the sun to shine or the wind to blow to suit our desires. It may be possible, to a limited degree, to make intermittent solar or wind energy *act like* fossil fuels by storing some of the electricity generated for later use, building extra capacity, or redesigning electricity grids. But this costs both money and energy. To avoid enormous overall system costs for capacity redundancy, energy storage, and multiple long-distance grid interconnections, it will be necessary to find more and more ways to shift electricity demand from times of convenience to times of abundant supply, and to significantly reduce overall demand.

2. *The liquid fuels problem.* As we will see in chapter 4, electricity doesn't supply all our current energy usage and is unlikely to do so in a renewable future. Our single largest source of energy is oil, which still fuels nearly all transportation as well as many industrial processes. While there are renewable replacements for some oil products (e.g., biofuels), these are in most cases not direct substitutes (few automobiles, trucks, ships, or airplanes can burn a pure biofuel without costly engine retrofitting) and have other substantial drawbacks and limitations.[4] Only portions of our transport infrastructure lend themselves easily to electrification—another potential substitution strategy. Thus a renewable future is likely to be characterized by less mobility, and this has significant implications for the entire economy.

3. *Other uses of fossil fuels.* Society currently uses the energy from fossil fuels for other essential purposes as well, including the production of high temperatures for making steel and other metals, cement, rubber, ceramics, glass, and other manufactured goods. Fossil fuels also serve as feedstocks for materials (e.g., plastics, chemicals, and pharmaceuticals). As we will see in chapter 5, all of these pose substitution or adaptation quandaries.

4. *Area density of energy collection activities.* In the energy transition, we will move from sources with a small geographic footprint (e.g., a natural gas well) toward ones with much larger footprints (large wind and solar farms collecting diffuse or ambient sources of energy). As we do, there will be unavoidable costs, inefficiencies, and environmental impacts resulting from the increasing spatial extent of energy collection activities. While the environmental impacts of a wind farm are substantially less than those from drilling for, distributing, and burning natural gas, or from mining, transporting, and burning coal, capturing renewable energy at the scale required to offset all gas and coal energy would nevertheless entail environmental impacts that are far from trivial. Minimizing these costs will require planning and adaptation.

5. *Location.* Sunlight, wind, hydropower, and biomass are more readily available in some places than others. Long-distance transmission entails significant investment costs and energy losses. Moreover, transporting biomass energy resources (e.g., biofuels or wood) reduces the overall energy profitability of their use. This implies that, as the energy transition accelerates, energy production will shift from large, centralized processing and distribution centers (e.g., a 500,000 barrel per day refinery) to distributed and smaller-scale facilities (e.g., a local or regional biofuel factory within a defined collection zone or "shed"), since the same amount of "feedstock" cannot be concentrated in one place. It also implies that population centers may tend to reorganize themselves geographically around available energy sources.

6. *Energy quantity.* As we will see in chapter 6, quantities of energy available will also change during the transition. Since the mid-nineteenth century, annual global energy consumption has grown exponentially to over 500 exajoules (fig. I.3). Even assuming a massive build-out of solar and wind capacity during the next 35 years, renewables will probably be unable to fully replace the quantity of energy currently provided by fossil fuels, let alone meet projected energy demand growth. This raises profound questions not only about how much energy will be available but also for widespread expectations and assumptions about global economic growth.

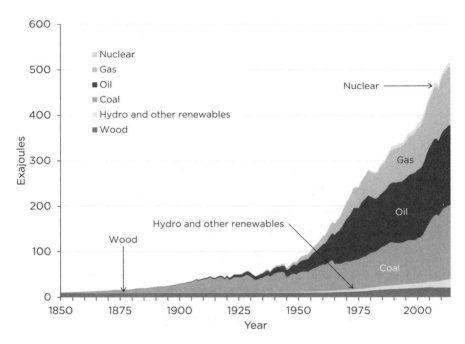

Figure I.3. World primary energy consumption by fuel type, 1850–2014. Primary electricity converted by direct equivalent method.
Source: Data compiled by J. David Hughes. Post-1965 data from BP, *Statistical Review of World Energy* (annual), http://bp.com/statisticalreview. Pre-1965 data from Arnulf Grubler, "Technology and Global Change: Data Appendix," (1998), http://www.iiasa.ac.at/~gruebler/Data/TechnologyAndGlobalChange/.

For these six reasons, we should explore now how energy *usage* must and will evolve during the next few decades as the world transitions (willingly or not) to renewable energy. As we've already seen, our current patterns of energy use developed in response to the qualities and quantities of the energy available to us during the past century. Fossil fuels provided significant advantages: they were available on demand, cheap, portable, and energy dense. They also entailed costs, including climate change and other environmental and social impacts.[5] Renewables offer their own suite of advantages, the most obvious of which are that, with solar and wind, there is no fuel cost, and they have far lower climate and health impacts. But that doesn't mean these are truly free or limitless energy sources: the devices used to capture energy from sunlight and wind require materials and embodied energy. Further, the energy we get from these sources is variable and won't substitute for all current fossil fuel uses. And the technologies we use to harvest energy from sunlight and wind have their own environmental impacts.

Engineers will certainly make every effort to adapt new energy resources to familiar usage patterns (e.g., by replacing gasoline-fueled cars with electric cars). We can, to a certain extent, press solar and wind into the mold of our current energy system by buffering their variability with energy storage technology and grid enhancements. But the larger the proportion of our total energy we get from these resources, the more our buffering efforts will cost in both money and energy. Over the long run, usage patterns will almost certainly change substantially as we adapt to renewable energy resources.

The problem with our current energy usage patterns is not simply that they are wasteful (though they often are) or that we use energy to do things that are harmful (though we often do). Even disregarding those legitimate concerns, many current energy usage patterns probably just won't work in an all-renewable world.

Overview of This Book

While the main thrust of this book is to examine how energy usage is likely to change in an all-renewable world, we will begin by reviewing the basics of energy and looking closely at how we currently power society.

Then we will take a survey of *energy supply and demand issues*, exploring the changed circumstances to which society will be adapting. This portion of the book consists of five chapters—three discussing energy quality (one on electricity, one on liquid fuels, one on other energy uses), one exploring how much renewable energy capacity might be available by midcentury, and one answering various objections likely to be raised to our conclusions about future energy supply.

The book concludes by discussing the critically important questions of how to ensure that *everyone* benefits from the renewable energy transition and what steps can and should be taken now to put us on a path toward a truly just and sustainable future.

The goal of this book is to help readers think more clearly and intelligently about our renewable future. An all-renewable world will present opportunities as well as challenges. And building that world will entail more than just the construction of enormous numbers of solar panels and wind turbines. Along the way, we will learn that how we *use* energy is as important as how we *get* it. Indeed, unless we adapt our energy *usage patterns* with the same vigor as is devoted to changing energy *sources*, the transition could result in a substantial reduction of economic functionality for society as a whole.

PART I

The Context:
It's All About Energy

CHAPTER 1

Energy 101

IT IS IMPOSSIBLE TO OVERSTATE the importance of energy. Without it, we can do literally nothing. Further, the unfolding consequences of modern civilization's energy use (including climate change), together with the inevitable energy transition from fossil fuels to renewables, will be the defining trends of the current century. How we address the climate–energy dilemma will make a life-or-death difference for current and future generations of humans, and for countless other species.

But we can't participate usefully in discussions about energy without some basic knowledge of the subject. This chapter surveys a few simple energy concepts that everyone should be familiar with. These will include the definition of "energy" and an exploration of the forms it takes; the difference between *energy* and *power*; the *Laws of Thermodynamics*; the distinction between *stocks* and *flows* of energy; *net energy* (or *energy returned on energy invested*); *lifecycle analysis* (LCA) and *lifecycle impacts*; and the difference between *operational energy* and *embodied energy*.

What Is Energy? The Basics of the Basics

Energy is known by what it does: physicists define it as the capacity to do work. Energy exists in several forms—including thermal, radioactive decay, kinetic, mechanical, and electrical—and its form can change. For example, the energy stored in the molecular structures of coal can be released as heat through the process of combustion. That heat can be used to boil water, creating steam at high pressure, which can flow through turbines that spin magnets to produce an electric current, which is then passed through transformers and wires into homes and offices, where it is available to power computers, lights, and televisions.

Energy is measured in a variety of units, including joules, British thermal units (BTUs), and calories. When discussing electrical energy, the most common unit is the *watt-hour* (Wh). Sometimes the energy of fossil fuels is discussed in terms of barrels-of-oil-equivalent (boe). Where renewable energy is used in the form of electricity, we will discuss it in terms of watt-hours and megawatt-hours (MWh).

It is also important to understand the distinction between energy and power. While units of energy measure the *total quantity* of work done, they don't tell us *how fast* that work is being accomplished. For example, you could lift a one-ton boulder up the side of a mountain using only a small electric motor and a system of pulleys, but it would take a long time. A more powerful electric motor could do the job faster, while a still more powerful rocket engine could rapidly propel a payload of identical weight to the top of the mountain in a matter of seconds. Power is therefore defined as the *rate at which energy is produced or used*. Think of it as energy per unit of time. The standard unit of electrical power is the watt (W). The amount of electrical energy a 10 W light bulb uses depends on how long it is lit: in one hour, it will use 10 Wh of energy. In the same amount of time, a hundred thousand such bulbs would use 1000 kilowatt-hours (kWh), which equals 1 MWh (1,000,000 watts = 1000 kilowatts = 1 megawatt).

Laws of Thermodynamics

Two important physical principles, known as the *first and second laws of thermodynamics*, describe limits to the ways that energy works.

The first, known as the *conservation law,* states that energy cannot be created or destroyed, only transferred or transformed. In the example cited earlier (the use of coal to power household appliances), the total amount of energy is conserved at every stage. When energy chemically stored in the coal was transformed into heat, then electric current, and finally into the work of lighting our office or running our computer, some was "lost" at each stage. But that energy still exists; it is merely released to the environment, mostly as heat.

That's where the second law, sometimes called the *entropy law,* comes in: it states that, whenever energy is converted from one form to another, at least some of it is dissipated (again, typically as heat). Though that dissipated energy still exists, it is now diffuse and scattered and thus not available to do work. Thus, in effect, usable energy is always being lost. The word *entropy* was coined by the German physicist Rudolf Clausius in 1868 as a measure of the amount of energy no longer practically capable of conversion into work. According to the second law, the entropy within an isolated system inevitably increases over time. Since it takes work to create and maintain order within a system, the entropy law tells us the depressing news that, in the battle between order and chaos, it is chaos that ultimately will win.

All of this means that it is technically incorrect to say that we "consume" (or "produce") energy. We merely obtain it from places of higher concentration and get it to do work for us before it dissipates into less concentrated forms (ultimately, low-level dispersed heat). Hence *energy density* is an important criterion in assessing the likely value of potential energy resources (see the section "Energy Resource Criteria" later in the chapter). Only relatively concentrated energy is useful to us. We live in a universe teeming with energy; the trick is putting that energy to work. This is much easier when some of that energy happens to be temporarily concentrated in fossil fuels, wood, or uranium, or in a persistent flow pattern (e.g., a river, wind, or sunlight).

Areas of higher energy concentration can take either of two forms: *stocks* or *flows.* A stock is a store of energy—energy chemically stored in wood, oil, natural gas, or coal, or nuclear energy stored in stocks of uranium. Flows of energy include rivers, wind, and sunlight. Both stocks and flows present challenges: flows tend to be variable, whereas stocks can be depleted.

Some forms of energy are more versatile in their usefulness than others. For example, we can use electricity for a myriad of applications, whereas the heat from burning coal is currently used mostly for stationary applications like generating power (we formerly burned it in locomotives and ships, until oil proved its superiority for mobile applications). When we turn the heat from burning coal into electricity, a substantial amount of energy is lost due to the inefficiency of the process. But we are willing to accept that loss because coal is relatively cheap, and it would be difficult and inconvenient to use burning coal *directly* to power lights, computers, and refrigerators. In effect, we put a differing value on different forms of energy, with electricity at the top of the value ladder, liquid and gaseous fuels in the middle, and coal or firewood at the bottom. Solar and wind technologies have an advantage in that they produce high-value electricity directly.

Energy efficiency can be defined as minimizing the loss of energy in the process of obtaining work from an energy source. (How far can we get a gallon of gas to propel a two-ton automobile? That is the car's fuel efficiency, which is often expressed in miles per gallon). Efficiency also applies to converting energy from one form to another. (How much electricity can a power plant generate from a ton of coal or a thousand cubic feet of natural gas? Power plant efficiency is usually expressed as a percentage, indicating how much of the initial energy is still available after conversion).

Net Energy

It takes energy to get energy: for example, energy is needed to drill an oil well or build a solar panel. Only *net energy*, what is left over after our energy investment is subtracted, is actually useful to us for end-use purposes. Sometimes the relationship between energy investments and yields is expressed as a ratio, *energy returned on energy invested* (EROEI, or sometimes just EROI). For example, an EROEI of 10:1 indicates ten units of energy returned for every unit invested.

The historic economic bonanza resulting from society's use of fossil fuels partly ensued from the fact that, in the twentieth century, only trivial amounts of energy were required in drilling for oil or mining for coal as compared to the gush of energy yielded. High EROEI ratios (in the range of 100:1 or more[1]) for society's energy-obtaining efforts meant that relatively small amounts of capital and labor were needed in order to supply all the energy that society could use. As a result, many people could be freed from basic energy-producing activities (like farming or forestry), their labor being substituted by fuel-fed machines. Channeled into manufacturing and managerial jobs, these people found ways to use abundant, cheap energy to produce ever more goods and services. The middle class mushroomed, as did cities and suburbs. In the process, we discovered an unintended consequence of having armies of cheap "energy slaves": as manufacturing and other sectors of the economy became mechanized, many handcraft professions disappeared.

The EROEI ratios for fossil fuels are declining[2] as the best-quality resources are used up; meanwhile, the net energy figures of most renewable energy sources are relatively low compared to fossil fuels in their heyday (and this is especially true when buffering technologies—such as storage equipment and redundant capacity—are accounted for[3]). A practical result of *declining* overall societal EROEI is the need to devote proportionally more capital and labor to energy production processes. This would likely translate, for example, to the requirement for more farm labor, and to fewer opportunities in professions not centered on directly productive activities: we would need more people making or growing things, and fewer people marketing, advertising, financing, regulating, and litigating them.[4] For folks who think we have way too much marketing, advertising, financialization, regulation, and litigation in our current society, this might not seem like such a bad thing.

Net energy analysis (NEA) establishes a baseline for the economic usefulness of any energy resource. Decades of research suggest that, if an energy resource cannot yield at least three units of energy for every unit employed in energy production, it will not be economically useful in the long run[5] unless the utilization of the energy is highly productive. A low net energy resource (such as biofuel) could potentially be of value if it provides a large benefit (e.g., as fuel for aviation if petroleum were to become scarce), but a high-EROEI resource would then be needed to provide the energy for the production of that lower-EROEI resource.

Unfortunately, the net energy or EROEI literature is inconsistent because researchers have so far been unable to agree on a common set of system boundaries. Therefore two analysts may calculate very different EROEI ratios for the same energy source. This does not entirely undermine the usefulness of NEA; it merely requires us to use caution in comparing the findings of different studies.

Incorporating the dimension of time into EROEI analysis adds yet another layer of complexity, but doing so is essential if we are to realistically compare energy from flows (solar and wind) with energy from stocks (fossil fuels). The great majority of the energy investment into solar panels comes during their manufacture, while the energy return is delivered slowly over the decades of their projected usefulness. This front-loading of energy investment creates problems if we wish to push the energy transition very quickly (as is documented in a study by Dale and Benson at Stanford University, who found that all solar technology installed until about 2010 was a net energy sink, in the sense that it hadn't yet paid for itself in energy terms[6]). If solar and wind build-out rates are very high, the net energy available from these sources will be smaller during the transition period (though wind's higher EROEI should result in shorter delays in system-wide energy payback) (fig. 1.1). During at least the

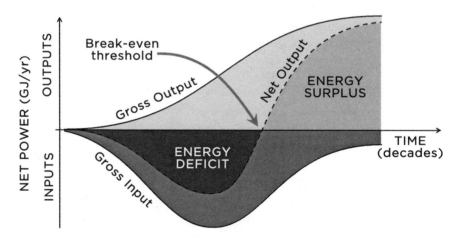

Figure 1.1. Energy input, output, and net power during the build-out of new energy production infrastructure. GJ = gigajoules.
Source: Michael Carbajales-Dale, "Fueling the Energy Transition: The Net Energy Perspective," presentation at Global Climate and Energy Project Workshop on Net Energy Analysis, Stanford University, Stanford, CA, April 1, 2015, https://gcep.stanford.edu/pdfs/events/workshops/Dale_GCEP_NEA_workshop_session_6-1.pdf.

early stages of the transition, the kinds of energy being invested in building and deploying renewable energy systems (mostly fossil fuels for high-heat industrial purposes and for transportation) will be different from the higher-quality electrical energy yielded from those systems.

Life Cycle Impacts

Life cycle analysis (LCA) assesses the resource burden and potential environmental impacts associated with a product, process, or service by compiling an inventory of relevant energy and material inputs and environmental releases, and by evaluating the potential environmental impacts associated with identified inputs and releases. LCA is used to evaluate not just energy technologies but products and services of all kinds. However, since it typically tracks energy inputs (among other things), it is highly relevant for understanding our current energy systems and for planning the transition to renewables.

Virtually all energy processes entail environmental impacts, but some have greater impacts than others. These may occur during the acquisition of an energy resource (as in mining coal), or during the release of energy from the resource (as in burning wood, coal, oil, or natural gas), or in the conversion of the energy from one form to another (as in converting the kinetic energy of flowing water into electricity via a dam and hydro-turbines).

Some environmental impacts are indirect and occur in the manufacturing of the equipment used in energy harvesting or conversion. For example, the extraction and manipulation of resources used in manufacturing solar panels entail significantly more environmental damage than the operation of the panels themselves.

NEA and LCA are complementary: NEA ignores the environmental costs of energy production activities, whereas LCA identifies and quantifies these; on the other hand, LCA tells us nothing about the economic viability of an energy technology. LCA can be narrowed to encompass just materials use or energy use, or broadened to include greenhouse gas emissions and other environmental impacts. A shortcoming of LCA studies is that they represent a snapshot in time; for a process that hasn't changed much in decades (e.g., cement making) this is not a big issue, whereas for communications technology it means that studies can quickly become outdated.

LCA also gives us useful information about our *energy-demand* activities—manufacturing, transportation, building operation and maintenance, food production, communication, health care, and so on.

Operational versus Embodied Energy

Another essential energy concept has to do with the difference between embodied and operational energy. When we estimate the energy required by an automobile, for example, we are likely to think at first only of the gasoline in its tank. However, a substantial amount of energy was expended in the car's construction, in the mining of ores from which its metal components were made, in the manufacture of the mining equipment, and so on. Further, enormous amounts of energy were spent in building the infrastructure that enables us to use the car—including systems of roads and highways, and networks of service stations, refineries, pipelines, and oil wells. The gasoline in the car's fuel tank supplies operational energy, but much more energy is embodied in the car itself and its support systems (fig. 1.2). This latter energy expenditure is easily overlooked.

The energy glut of the twentieth century enabled us to embody energy in a mind-numbing array of buildings, roads, pipelines, machines, gadgets, and packaging. Middle-class families got used to buying and discarding enormous quantities of manufactured goods representing generous portions of previously expended energy. If we have less energy available to us in our renewable future, this may impact more than the operation of our machines and the lighting and heating of our buildings. It means those buildings, that infrastructure, and all those manufactured goods will be increasingly expensive to produce, which may translate to a shrinking flow of manufactured goods that embody past energy expenditure, and a reduced ability to construct high-energy-input structures. We might have to get used to consuming simpler foods again, rather than highly processed and excessively packaged ones. We might purchase, on average, fewer items of clothing and furniture and fewer electronic devices, and

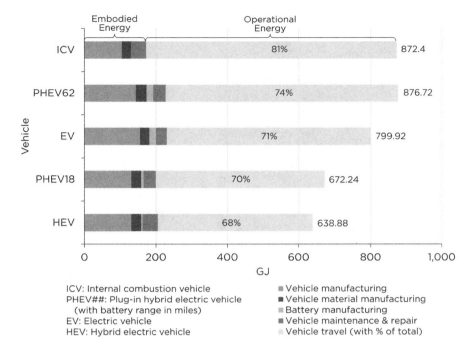

Figure 1.2. Life cycle embodied and operational energy of five vehicle types. Electricity use in EVs and PHEVs is based on the U.S. power supply mix.
Source: Nuri Cihat Onat, Murat Kucukvar, and Omer Tatari, "Conventional, Hybrid, Plug-in Hybrid or Electric Vehicles? State-Based Comparative Carbon and Energy Footprint Analysis in the United States," *Applied Energy* 150 (2015): 36–49, doi:10.1016/j.apenergy.2015.04.001.

we might inhabit smaller spaces. We might use old goods longer, and reuse and repurpose whatever can be repaired. Exactly how far these trends might proceed is impossible to say. Nevertheless, under such conditions it is fair to assume that this overall shift might constitute the end of the network of economic arrangements collectively known as *consumerism*.[7] Here again, there are more than a few people who believe that advanced industrial nations consume excessively, and that some simplification of upper- and middle-class lifestyles would be a good thing.

The distinction between operational and embodied energy is important in the context of this report because it may well turn out to be much easier to operate machines and systems with renewable energy than to embody renewable energy into materials, machines, and infrastructure.

Energy Resource Criteria

In evaluating energy sources, NEA and LCA are essential, but the following criteria also need to be taken into account:

1. Energy density
2. Direct monetary cost
3. Other resources needed
4. Renewability
5. Scalability
6. Location
7. Reliability
8. Transportability

1. ***Energy density.*** Measures of energy density include mass density, volume density, and area density.
   a. *Mass (or gravimetric) density.* Mass density is the amount of energy contained per unit of mass of an energy resource. For example, if we use the megajoule (MJ) as a measure of energy and the kilogram (kg) as a measure of mass, coal has about 20 to 35 MJ/kg, natural gas about 55 MJ/kg, and oil around 42 MJ/kg (for comparison's sake, the amount of food that a typical diet-conscious American eats throughout the day weighs a little over a kilogram (dry) and has an energy value of about 10 MJ, or 2,400 kilocalories[8]). However, an electric battery is typically able to store and deliver only about 0.1 to 0.5 MJ/kg, and this is why electric batteries are problematic in transport applications: they are very heavy in relation to their energy output. Thus electric cars tend to have limited driving ranges and electric aircraft (which are exceedingly rare) are able to carry only one or two people.

      Consumers and producers are sometimes willing to pay a premium for energy resources with a higher energy density by mass; therefore in some instances it might theoretically make economic sense to convert a lower-density fuel, such as coal, into a higher-density fuel, such as synthetic diesel, though the conversion process entails such high monetary and energy costs that most commercial efforts to do this have failed.

b. *Volume (or volumetric) density.* Volume density is the amount of energy contained within a given volume unit of an energy resource (e.g., MJ per liter [L]). Obviously, gaseous fuels will tend to have lower volumetric energy density than solid or liquid fuels. Natural gas has about .035 MJ/L at sea level atmospheric pressure, and 6.2 MJ/L when pressurized to 200 atmospheres. Oil, though, contains about 35 MJ/L.[9]

In most instances mass density is more important than volume density; however, for certain applications the latter can be decisive. For example, fueling airliners with hydrogen, which has high energy density by weight, would be problematic because it is a diffuse gas at common temperatures and surface atmospheric pressure; thus a hydrogen airliner would require very large tanks (themselves having large mass) even if the hydrogen were supercooled and highly pressurized.

The greater ease of transporting a fuel of higher volume density is reflected in the fact that oil moved by tanker is traded globally in large quantities, while the global tanker trade in natural gas is relatively small. Consumers and producers will usually pay a premium for energy resources of higher volumetric density.

c. *Area density.* The area density is the amount of energy that can be obtained from a given land area (e.g., an acre or a hectare) when the energy resource is in its original state. For example, the area energy density of wood as it grows in a forest is roughly 1 to 5 million MJ per acre.[10] The area density for oil is usually tens or hundreds of millions of MJ per acre where it occurs, though oilfields are much rarer than forests (except perhaps in Saudi Arabia).

Area energy density matters because energy sources that are already highly concentrated in their original form generally require less investment and effort to be put to use. Thus energy producers often tend to prefer energy resources that have high area density, such as oil that can be refined into gasoline, over ones that are more widely dispersed, such as corn that is intended to make ethanol (fig. 1.3).

2. **Direct monetary cost.** This is the criterion to which most attention is normally paid. Clearly, energy must be affordable and competitively priced if it is to be useful to society. However, the monetary cost of energy does not always reflect its true cost, as some energy resources may benefit from hidden subsidies or may have costs that are not currently directly paid for by the buyer (called *external* costs, such as health or environmental impacts). The monetary cost of energy resources is largely

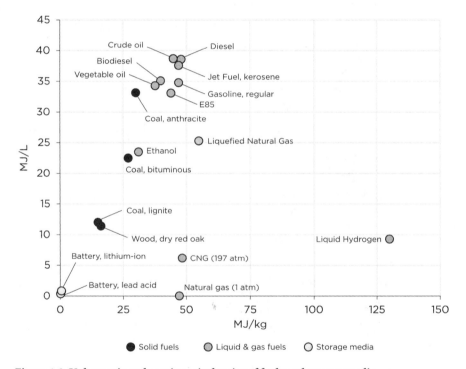

Figure 1.3. Volumetric and gravimetric density of fuels and storage media.
Sources: Coal: Tadeusz Patzek and Gregory Croft, "A Global Coal Production Forecast with Multi-Hubbert Cycle Analysis," *Energy* 35 (2010): 3111. Natural gas: https://people.hofstra .edu/geotrans/eng/ch8en/conc8en/energycontent.html. Crude oil and wood: http: //w.astro.berkeley.edu/~wright/fuel_energy.html. Batteries: http://www.allaboutbatteries. com/Battery-Energy.html and http://thebulletin.org/limits-energy-storage-technology. All others: Charles Hall and Kent Klitgaard, *Energy and the Wealth of Nations: Understanding the Biophysical Economy* (New York: Springer, 2012).

determined by the other criteria listed here, as well as supply and demand (table 1.1).

3. *Other resources needed.* Very few energy sources come in an immediately useable form. We can be warmed by the sunlight that falls on our shoulders on a spring day without exerting effort or employing any technology. But most energy resources, in order to be useful, require a method of gathering and/or converting the energy. This usually entails the use of some kind of apparatus, made of some kind of material (e.g., oil-drilling equipment is made from steel and the bits from diamonds). The extraction or conversion process generally also uses some kind of energy resource (e.g., the production of synthetic diesel fuel from tar sands requires water and heat; natural gas is often used). The availability or scarcity of the material or resource, and the complexity and cost of the apparatus, thus constitute limiting factors on energy production.

Table 1.1. U.S. average estimated levelized cost of electricity (LCOE) for new plants entering service in 2020 ($[2013]/MWh). O&M = operations and maintenance.

Plant Type	Capacity Factor (%)	Levelized Cost	O&M, fuel, and transmission investment	With available tax credits	Total LCOE incl. tax credits
Dispatchable Technologies					
Conventional Coal	85	60.4	34.8		95.1
Advanced Coal	85	76.9	38.8		115.7
Advanced Coal with CCS	85	97.3	47.1		144.4
Natural Gas-fired					
Conventional Combined Cycle	87	14.4	60.7		75.2
Advanced Combined Cycle	87	15.9	56.8		72.6
Advanced CC with CCS	87	30.1	70.1		100.2
Conventional Combustion Turbine	30	40.7	100.9		141.5
Advanced Combustion Turbine	30	27.8	85.8		113.5
Advanced Nuclear (online in 2022)	90	70.1	25.1		95.2
Geothermal	92	34.1	13.7	-3.4	44.4
Biomass	83	47.1	53.3		100.5
Non-Dispatchable Technologies					
Wind	36	57.7	15.9		73.6
Wind – Offshore	38	168.6	28.3		196.9
Solar PV	25	109.8	15.5	-11.0	114.3
Solar Thermal	20	191.6	48.1	-19.2	220.6
Hydroelectric	54	70.7	12.9		83.5

Source: U.S. Energy Information Administration, *Levelized Cost and Levelized Avoided Cost of New Generation Resources in the Annual Energy Outlook 2015*, June 2015; https://www.eia.gov/forecasts/aeo/pdf/electricity_generation.pdf.

The requirements for ancillary resources in order to produce a given quantity of energy are usually reflected in the price paid for the energy. But this is not always or entirely the case. For example, some thin-film photovoltaic (PV) panels incorporate materials such as gallium and indium that are nonrenewable, rare, and depleting quickly.[11] While the price of thin-film PV panels reflects the current market price of these materials, it does not give an indication of future limits to the scaling up of thin-film PV due to these materials' increasing scarcity.

4. *Renewability.* If we wish our society to continue using energy at industrial rates of flow not just for years or even decades but for centuries into the future, then we will require energy sources that can be sustained more or less indefinitely. Energy resources like oil, natural gas, and coal are clearly nonrenewable because the time required to form them through natural processes is measured in the tens of millions of years, whereas the stocks available will power society at best for only a few decades into the future at current rates of use. In contrast, solar PV and solar thermal energy sources rely on sunlight, which for practical purposes is not depleting and will presumably be available in similar quantities a thousand years hence.

Some energy resources are renewable yet are still capable of being depleted. For example, wood can be harvested from forests that regenerate themselves. However, the rate of harvest is crucial: if overharvested, the trees will be unable to regrow quickly enough and the forest will shrink and disappear. Even energy resources that are renewable and that do not suffer depletion are nevertheless limited by the size of the resource base (as will be discussed next).

5. *Scalability.* Estimating the potential contribution of an energy resource is obviously essential for macroplanning purposes, but such estimates are always subject to error—which can sometimes be significant. With fossil fuels, amounts that can be reasonably expected to be extracted and used on the basis of current extraction technologies and fuel prices are classified as *reserves*, which are always a fraction of *resources* (defined as the total amount of the substance estimated to be present in the

ground). For example, the U.S. Geological Survey's first estimate of national coal reserves, completed in 1907, identified 5000 years' worth of supplies. In the decades since, most of those reserves have been reclassified as resources, so that today only 250 years' worth of U.S. coal supplies are officially estimated to exist—a figure that is still probably much too optimistic (as the National Academy of Sciences concluded in its 2007 report, *Coal: Research and Development to Support National Energy Policy*[12]). Reserves are downgraded to resources when new limiting factors are taken into account, such as, in the case of coal, seam thickness and depth, chemical impurities, and location of the deposit.

On the other hand, reserves can sometimes grow as a result of the development of new extraction technologies, as has occurred in recent years with U.S. natural gas supplies: while the production of conventional American natural gas is declining, new horizontal drilling and underground fracturing technologies have enabled the recovery of "unconventional" gas from low-porosity rock (usually shale), significantly increasing the national natural gas production rate and expanding U.S. gas reserves.

Reserves estimation is especially difficult when dealing with energy resources that have little or no extraction history. This is the case, for example, with methane hydrates, with regard to which various experts have issued a very wide range of estimates of both total resources and extractable future supplies; it is also true of oil shale, also known as kerogen-rich marlstone (not to be confused with tight oil, which is sometimes confusingly called shale oil), and to a lesser degree tar sands, all of which have limited extraction histories.

Estimating potential supplies of renewable resources such as solar and wind power is likewise problematic, as many limiting factors are often initially overlooked. With regard to solar power, for example, a cursory examination of the ultimate potential is highly encouraging: the total amount of energy absorbed by Earth's atmosphere, oceans, and land masses from sunlight annually is approximately 3,850,000 exajoules (EJ)—whereas the world's human population currently uses just over 500 EJ of energy per year from all sources combined,[13] an insignificant fraction of the previous figure. However, the factors limiting the amount of sunlight that can potentially be put to work for humanity are numerous, starting with the material and land use requirements for the building and siting of solar collectors. The technical potential for global solar power is calculated by various researchers at between 43 and 2592 EJ—and for wind, between 72 and 700 EJ—according to figures collected by Moriarty and Honnery.[14]

Or consider the case of methane harvested from municipal landfills. In this instance, using the resource provides an environmental benefit: methane is a more powerful greenhouse gas than carbon dioxide, so harvesting and burning landfill gas (rather than letting it diffuse into the atmosphere) reduces climate impacts while also providing a local source of fuel. If landfill gas could power the U.S. electrical grid, then the nation could cease mining and burning coal. However, the potential size of the landfill gas resource is woefully insufficient to support this. Currently the nation derives about 16 billion kWh per year from landfill gas for commercial, industrial, and electric utility uses. This figure could probably be doubled if more landfills were tapped.[15] But U.S. electricity consumers use close to 200 times as much energy as that. Landfills deplete like oil wells. Further, more modern rubbish consists so much of paper and plastic that it probably won't produce methane in quantities that are useful; the methane being harvested now is largely from trash buried in the 1970s when a greater proportion of rubbish was putrescible. There is still another wrinkle: if society were to become more environmentally prudent and energy efficient, the result would be that the amount of trash going into landfills would decline—and this would reduce the amount of energy that could be harvested from future landfills.

6. *Location.* The fossil fuel industry has long faced the problem of "stranded gas"—natural gas reservoirs that exist far from pipelines and that are too small to justify building pipelines to access them. Renewable resources often face similar hurdles.

The location of solar and wind installations is largely dictated by the availability of the primary energy resource; often, sun and wind are most abundant in sparsely populated areas. For example, in the United States there is large potential for the development of wind resources in Montana and North and South Dakota; however, these are three of the least-populous states in the nation. Therefore, to take full advantage of these resources it would be necessary to build high-capacity power lines from these states to more populated regions. There are also good wind resources offshore along the Atlantic and Pacific coasts, nearer to large urban centers, but taking advantage of these resources will entail overcoming challenges having to do with building and operating turbines in deep water and connecting them to the grid onshore. Similarly, the nation's best solar resources are located in the Southwest, far from the population centers of the East Coast and Midwest.

Thus taking full advantage of these energy resources would require more than the construction of wind turbines and solar panels: much of the U.S. electricity grid would need to be reconfigured, and large-capacity, long-distance transmission lines would need to be constructed.

7. *Reliability.* Some energy resources are continuous: coal can be fed into a boiler at whatever rate the technology is able to accommodate, as long as the coal is available. But some energy sources, such as wind and solar, are subject to rapid and unpredictable fluctuations. Another way to say this is that our rate of using some resources is capacity constrained (e.g., by the size of the boiler and conveyor belts), while wind and solar are supply constrained, since the availability of the resource (sunshine or wind) dictates the rate of delivery. The wind often blows at greatest intensity at night, when electricity demand is lowest. The sun shines for the fewest hours per day during the winter—but power system operators are required to assure security of supply throughout the day and year (fig. 1.4).

Monthly Production Solar and Wind

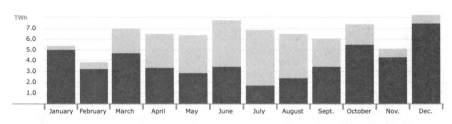

Weekly Production Solar and Wind

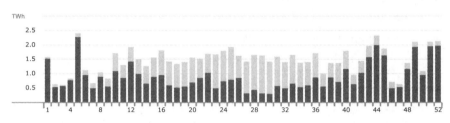

Daily Production Solar and Wind

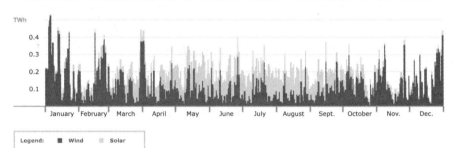

Legend: ■ Wind   ▨ Solar

Figure 1.4. Monthly, weekly, and daily electricity production of solar and wind power in Germany, 2013.
Source: Bruno Burger, "Electricity Production from Solar and Wind in Germany in 2013" (Freiburg: Fraunhofer ISE, January 9, 2014), https://www.ise.fraunhofer.de/en/renewable-energy-data.

As noted previously, intermittency of energy supply can be managed to a certain extent through storage systems, capacity redundancy, and grid upgrades. However, these imply extra infrastructure costs as well as energy losses.

8. ***Transportability.*** The transportability of energy is largely determined by the mass and volume density of the energy resource, as already discussed. But it is also affected by the state of the material (assuming that

it is a substance)—whether it is a solid, liquid, or gas. In general, a solid fuel is less convenient to transport than a gaseous fuel because the latter can move by pipeline (pipes can move eight times the volume on a doubling of the size). Liquids are the most convenient of all because they can likewise move through hoses and pipes, take up less space than gases, and won't dissipate into the air if not stored and distributed in perfectly airtight systems.

Energy resources that are fluxes or flows, like the energy from sunlight or wind, cannot be directly transported; they must first be converted into a form that can be—such as electricity. Electricity is highly transportable because it moves through wires, enabling it to be delivered not only to nearly every building in industrialized nations but to many locations within each building.

Transporting energy always entails costs—whether it is the cost of hauling coal (which may account for up to 70 percent of the delivered price of the fuel[16]), the cost of building and maintaining pipelines and pumping oil or gas, or the cost of building and maintaining an electricity grid. Using the grid entails costs too, since energy is lost in transmission (about 6 percent in the United States[17]). These costs can be expressed in monetary terms or in energy terms. The energy costs of transporting energy affect net energy.

Obviously, the evaluation of energy resources is a complicated process that entails the likelihood of estimation errors. The failure to take a single challenging criterion into account can lead to unwarranted optimism regarding the promise of an energy resource; on the other hand, the failure to foresee technological innovation can lead to too much pessimism. Some energy analysts advocate letting the market sort out energy winners and losers, but that's not an adequate response to the challenges presented by climate change and fossil fuel depletion: if we wait for the market to force an energy transition, it will do so too slowly and too late to prevent both environmental and economic turmoil. Further, the market tends to look for short-term results; therefore, market-driven solutions may not be sustainable long term.

CHAPTER 2

A Quick Look at Our Current Energy System

THE STATISTICS ARE READILY AVAILABLE: our world presently uses about 520 quadrillion British thermal units each year, or 153 billion megawatt-hours—the equivalent of 100 billion barrels of oil.[1] These numbers are readily interpreted by the experts. But what do they mean in terms the nonspecialist can understand?

A hard-working human can generate power in the range of 60 to 300 watts,[2] depending on the person's strength and which muscles are in use. Since the upper part of that range is realistic only for trained athletes using their leg muscles, let's start with a more conservative and realistic number—100 watts. Sustained for an hour, that would be 100 watt-hours of energy. Working eight-hour days five days a week for a year, with no holidays, our hypothetical hard worker would produce 208,000 watt-hours of useful work, or 208 kilowatt-hours.

World annual energy usage therefore equals the energy output of the yearly manual labor of 734.4 billion humans—a hundred times the current global population (though a large portion of that energy is wasted). We have obviously come very far from the days, just a couple of centuries ago, when a quarter of all agricultural land was set aside to grow food for draft animals; when we derived our heat from burning wood; when most of the motive force in the economy derived from human and animal muscles; and when many human beings were enslaved so that their muscle power could be forcibly directed by other humans.

As we will see shortly, most of our current "energy slaves" are fossil-fueled, and their work is done mostly to the advantage of people in wealthy countries, whereas the poorest humans still get by largely on muscle power.

Growth

The single trend that best captures the history of energy use since the start of the industrial revolution is *growth*. Since 1850 (when world population stood at less than 1.3 billion), total yearly energy use has grown from about 10 exajoules per year to over 500 exajoules per year (see fig. I.3). Since 1980, population has grown from 4.4 billion to 7.3 billion in 2015, while total energy use has nearly doubled. On a per capita basis, supply of energy has grown since 1850 by nearly 900 percent (though in the past 40 years per capita growth has slowed; fig. 2.1). Since 1980 per capita energy use has increased nearly 20 percent, from 67 gigajoules (GJ) to about 80 GJ per year.

Clearly, growth has been occurring in both *energy* and *population*. Has energy growth caused population growth? Not directly: countries with high rates of energy use generally do not have high population growth rates, and most countries with very high population growth rates use relatively little energy on a per capita basis. However, advances in agriculture and public health that are directly and indirectly tied to energy growth have made possible a dramatic increase in population over time.

Growth in *energy* and *gross domestic product (GDP)* are also tied. Energy enables the activities that generate GDP, so the relationship is direct, but it is not static; we have gradually become more efficient in the use of energy in creating GDP (fig. 2.2). During the renewable energy transition we will be challenged to become more efficient still (we'll discuss the relationship between energy use

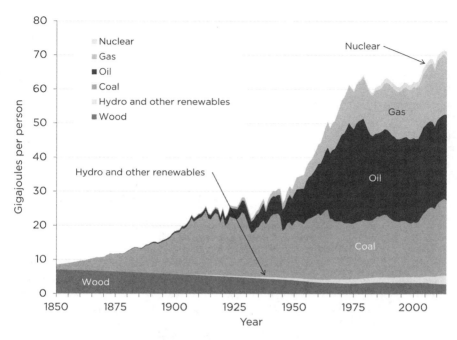

Figure 2.1. World per capita primary energy consumption per year by fuel type, 1850–2014. Primary electricity converted by direct equivalent method.
Source: Data compiled by J. David Hughes from Arnulf Grubler, "Technology and Global Change: Data Appendix," (1998), http://www.iiasa.ac.at/~gruebler/Data/Technology AndGlobalChange/ and BP, *Statistical Review of World Energy*, (annual) http://bp.com /statisticalreview.

and GDP further in chapter 6, "Energy Supply," under the heading "Energy Intensity"). Nevertheless, even if the energy–GDP linkage is stretchable, it is in the end unbreakable: it takes energy to do anything whatsoever.

Energy Rich, Energy Poor

Clearly, people in some countries use a lot more energy than people in others. From a human perspective, having little energy available means spending a lot of time in mundane activities related to daily life (cooking, washing clothes, walking, planting, weeding and harvesting, etc.). Having lots of energy means having machines do many of these things, or help do them; it also usually implies the ability to accelerate the pace of life and thus consume more goods,

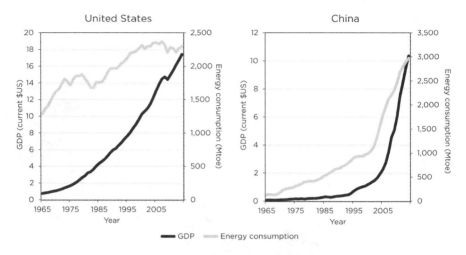

Figure 2.2. Total gross domestic product and energy consumption of United States and China, 1965–2014. The link between GDP and energy consumption is clear in countries in both later stages of economic development (e.g., United States) and earlier stages of economic development (e.g., China). Although the strength of the link appears to decline in the United States in later decades, this is in part because so much energy consumption for manufacturing has been relocated to other countries—including China.
Source: World Bank, World Development Indicators, http://data.worldbank.org/data
-catalog/world-development-indicators and BP, *Statistical Review of World Energy* (annual),
http://bp.com/statisticalreview.

have more experiences, travel farther and more often, and get an education and ultimately a better-paying, more highly skilled job. Spending money on such energy-consuming activities contributes significantly to GDP.

There is an obvious connection between energy inequality and economic inequality: very low energy use is associated with poverty, very high energy use with wealth (fig. 2.3). However, the connection is not absolute: for example, Germans enjoy a high standard of living, yet use only a little more than half as much energy (per capita) as citizens of the United States and Canada (fig. 2.4).

The fossil fuel era has produced great wealth, and some have partaken of that wealth far more than others. However, as we will discuss in more detail in chapter 8, "Energy and Justice," the end of the fossil fuel era does not necessarily imply the end of energy inequality. Solar panels and wind turbines require investment and produce benefits; in the renewable era ahead, it is certainly possible to imagine scenarios in which only some can afford the needed investment and can therefore enjoy the benefits. The degree to which energy inequality

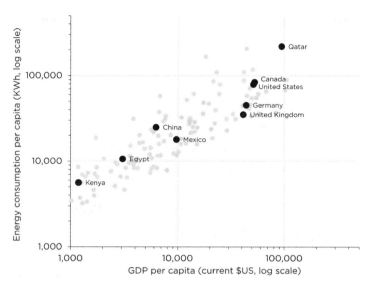

Figure 2.3. Per capita gross domestic product and energy consumption of various countries, 2012.
Source: World Bank, World Development Indicators, http://data.worldbank.org/data-catalog/world-development-indicators.

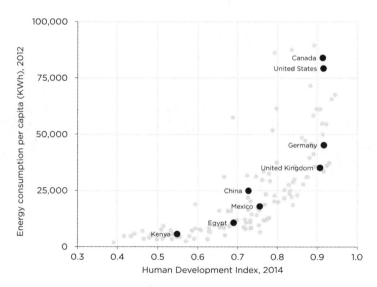

Figure 2.4. Human Development Index (2014) and per capita energy consumption (2012) for various countries.
Source: World Bank, World Development Indicators, http://data.worldbank.org/data-catalog/world-development-indicators and United Nations Development Program, http://hdr.undp.org/en/composite/HDI.

is either reduced or cemented into place will depend on how the transition is planned and implemented.

Energy Resources

These charts more or less speak for themselves (fig. 2.5). We currently draw upon many different energy resources, but just a few supply the bulk of all energy used: about 85 percent of our energy comes from oil, coal, and natural gas.

One factor is not readily apparent in the charts: in poor nations, a lot of energy comes from traditional biomass, such as burning wood, crop residues, and dung.

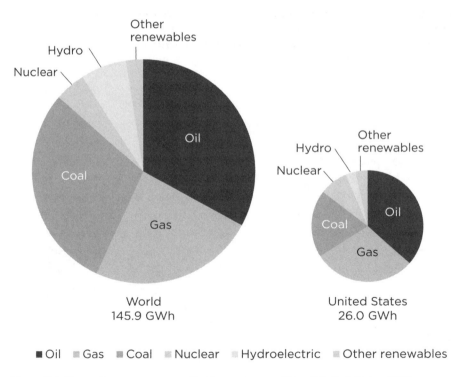

Figure 2.5. Current energy consumption by source, world and United States, 2012.
Source: BP, *Statistical Review of World Energy* (annual), http://bp.com/statisticalreview.

End Use

Primary energy is energy in its initial form, as it is directly extracted from Earth (crude oil, natural gas, coal) or as it is available without conversion through combustion (electricity from wind, solar, hydro, nuclear, tidal, etc.). *Final energy* is the energy we use directly in forms that are suited to their use (electricity for lighting, gasoline for internal combustion engines, kerosene for jet turbines, coke for steel making, etc.). In between the *primary* and *final* stages much energy is typically lost.

In 2012, primary energy production was 13,350 million metric tons of oil equivalent (Mtoe) while final consumption was 9100 Mtoe[3]—68 percent of what we started with, the rest having been lost both in conversion of primary energy to the forms we prefer to use (gasoline, electricity) and in use by the energy conversion industries themselves (fig. 2.6). The bulk of this conversion loss occurred in making electricity (average 54 percent losses globally)[4]—which demonstrates both the degree to which we value electricity and the easy

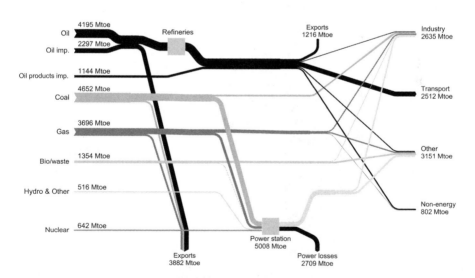

Figure 2.6. Flow of world energy production and consumption, 2012. Flows under 100 Mtoe and some imports not shown for clarity.
Source: International Energy Agency, "World: Balance (2012)," http://www.iea.org/sankey/#?c=World&s=Balance.

availability of fossil fuels to accommodate such a high proportion of losses. As
we have already noted more than once, this implies some good news for the renewable energy transition because wind and solar electricity do not entail these
conversion losses. There are also end-use losses, notably in the transportation
sector, due to the inefficiency of internal combustion engines in transforming
the energy stored in fuels into motive force.

We use energy in everything we do, so it is difficult to adequately capture
all the ways we use energy in a single chart. We can divide energy usage into sectors, for the United States and the world. It's helpful then to pick apart the energy use within each of those sectors. Let's look at the food system, for example,
in figure 2.7.

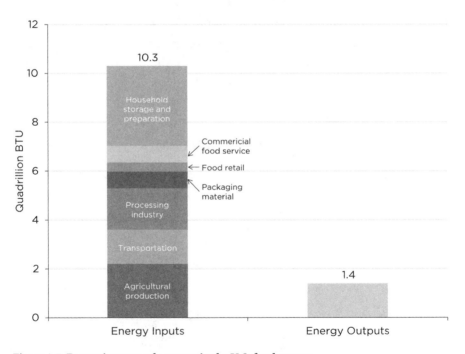

Figure 2.7. Energy inputs and outputs in the U.S. food system.
Source: Center for Sustainable Systems, University of Michigan, "U.S. Food System Fact-
sheet." Pub. No. CSS01-06 (2015), http://css.snre.umich.edu/css_doc/CSS01-06.pdf.

Changing our energy system will require both detailed and systemic thinking. Some aspects of the food system won't pose too big a challenge: we can use electricity from renewable sources to run existing machinery, while finding ways to cool food more efficiently and reduce the need for refrigeration. But other aspects will prove difficult to transition: tractors, combines, heavy trucks, and many other vehicles and machines are all currently built for fossil fuels and oriented on the global fossil fuel supply network. (We will discuss the challenges and implications of this transition in chapters 4 and 5.)

*   *   *

Let's summarize what we have learned so far. Energy is important, we use a lot of it, and we are in the very early stages of a great transition from overwhelming reliance on fossil fuels toward reliance on renewable sources. We've seen what energy is and how it works, as well as the criteria and tools to use in evaluating energy sources. We've explored the difference between operational and embodied energy, and between primary and final energy. We've also seen how unequally we consume energy, and what we use it for.

Now we are ready to explore some of the opportunities and challenges we may face during the transition.

PART II

Energy Supply
in a Renewable World:
Opportunities and Challenges

CHAPTER 3

Renewable Electricity: Falling Costs, Variability, and Scaling Challenges

THE UNIVERSAL AVAILABILITY and use of electricity has come to define modern life, at least for the vast majority of people in the industrialized world. Electricity is accessible in nearly every home and commercial building. We rely on power from wall sockets 24 hours a day, 365 days a year for a myriad of uses that range from toasting a bagel to powering an MRI machine. Electricity is remarkably versatile, and we have built a massive infrastructure to generate, distribute, and consume it.

Electricity constitutes only a portion of the energy the world uses daily. In the United States, 21 percent of final energy is used as electricity (for the world, the figure is 18 percent); of the U.S. electricity supply, 38 percent is generated from coal, 31 percent from natural gas, 19 percent from nuclear power, 7 percent from hydro, and 5 percent from other renewables (fig. 3.1).[1]

Since most solar and wind energy technologies produce electricity (as do hydro, geothermal, and some biomass generators), replacement of fossil fuels by renewable energy sources is happening fastest in the electricity sector. Fur-

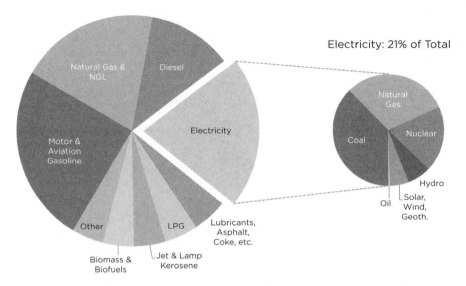

Figure 3.1. U.S. final energy consumption by fuel type, 2012. NGL = natural gas liquids; LPG = liquefied petroleum gas.
Source: International Energy Agency and U.S. Energy Information Administration.

ther, this means that hopes for accelerating the energy transition hinge on the electrification of a greater proportion of our total energy use.

For proponents of renewable energy, there has been plenty of good news in recent years regarding falling prices for solar and wind, and soaring growth rates in these industries. Still, as we will see in this chapter, there are significant challenges to be addressed.

Price Is Less of a Barrier

Solar and wind are growing fast. In 2014, global solar capacity grew 28.7 percent over the previous year and has more than quadrupled in the past four years.[2] This is an astounding rate of growth: if it were to continue, solar would become the world's dominant source of electricity by 2024. Wind energy capacity is growing at a somewhat slower pace (doubling about every five years), but has a larger current base: in 2012 (the last full year of U.S. Energy Information Administration [EIA] global data by generation type), solar delivered 94 terawatt-hours (TWh) (billion megawatt-hours [MWh]) per year, versus wind's 522 TWh per year, out of a global generation of 22,600 TWh.[3]

Remarkably, in the United States solar and wind power are currently growing faster than coal—not just in percentage terms but in absolute numbers: for 2014, the U.S. increase in coal consumption amounted to 4.6 TWh, while solar and wind added 23 TWh.[4] Even in China, solar and wind are expanding quickly, while coal consumption is hardly growing at all or even starting to taper off (owing to a substantial slowdown in industrial consumption).

Solar and wind's spectacular growth is occurring for several reasons, but perhaps the most significant driver has been the fall in prices for new solar and wind capacity as compared to costs for coal and natural gas. The price drop is most apparent in the case of solar: the price of photovoltaic (PV) cells has fallen by 99 percent over the past twenty-five years, and the trend continues. In a 2014 report, Deutsche Bank solar industry analyst Vishal Shah forecast that solar will reach "grid parity" in 36 of 50 U.S. states by 2016, and in most of the world by 2017 (grid parity is defined as the point where the price for PV electricity is competitive with the retail price for grid power).[5] Shah also estimates that installed solar capacity will grow as much as sixfold before the end of the decade; see figure 3.2 for a snapshot of just the last few years of solar capacity growth.

The fall in PV prices is being driven by two factors: improvements in technology (both in manufacturing methods and in PV materials), and increased scale of manufacturing. Manufacturing scale improvements have resulted largely from the Chinese government's decision in 2009 to support widespread deployment of PV, which in turn has led to a spate of price cutting across the industry, as well as a global flood of cheap panels—though some characterize China's actions as product dumping or unfair competition, with many American and European manufacturers having gone bankrupt due to their inability to match Chinese prices.

Power purchase agreement prices for wind energy projects are currently competitive with prices for power from coal and natural gas plants in many markets. Wind prices are falling because of lower-cost wind turbines (taller wind towers and longer and lighter blades) that allow for a better capture of the wind resource, and therefore increased economic performance.

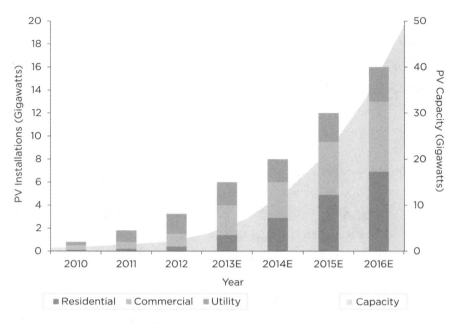

Figure 3.2. U.S. total photovoltaic installations and capacity.
Source: Vishal Shah, Jerimiah Booream-Phelps, and Susie Min, "2014 Outlook: Let the Second Gold Rush Begin," Deutsche Bank, Market Research, North America United States, January 6, 2014, https://www.deutschebank.nl/nl/docs/Solar_-_2014_Outlook_Let_the _Second_Gold_Rush_Begin.pdf.

In general, technologies tend to become more efficient and more cost-effective over time, as engineers identify improvements and as devices are produced on a larger scale.[6] Fossil fuel technologies (mining, drilling, hydrofracturing, refining) are also becoming more efficient; however, those technologies are being used to harvest depleting resources, so an accelerating decline in resource quality will inevitably outstrip the ability of engineers to improve recovery efficiencies.[7]

Is the current rapid growth in solar and wind capacity sustainable? Can the pace in fact be substantially increased? Will price declines increase or reverse themselves as higher penetration rates are achieved? The answers to these questions will depend on the renewable energy industry's ability to solve a few looming problems.

Intermittency

As stated earlier, we have designed our energy usage patterns to take advantage of controllable inputs. Need more electricity? If you're relying on coal for energy, that just requires shoveling more fuel into the boiler. Sunlight and wind are different: they are available on Nature's terms, not ours. Sometimes the sun is shining or the wind is blowing, sometimes not. Energy geeks have a vocabulary to describe this—they say solar and wind power are *intermittent, variable, stochastic,* or *chaotic.* In contrast, energy experts refer to coal, gas, oil, hydro, biomass, nuclear, and geothermal sources as *predictable*; sources that can be quickly brought into service or shuttered to meet transient demand (usually natural gas or hydro plants) are called *dispatchable.* It should be noted, though, that these latter sources are also subject to a certain amount of variability: natural gas, coal, and nuclear power plants sometimes need to be shut down for maintenance, or can go offline due to accidents, and hydropower can be distinctly seasonal depending on rainfall patterns. They're just much *less* variable than solar and wind.

The availability of sunlight follows fairly consistent diurnal and seasonal patterns. We can calculate in advance the position of the sun in the sky for any moment in time, for any location. We know that sunlight will be more readily available in summer months than in winter months, and that this seasonal variability will be more extreme the farther we are from the equator. We also know that sunlight is likely to be most intense at noon and is absent at night. Yet within this expected variability there is also a more chaotic intermittency: sometimes the sun is hidden for moments, hours, days, or even weeks by clouds.

Wind tends to follow different diurnal and seasonal patterns. Some locations have far more consistent winds than others. Also, winds tend to be stronger, and more consistent, at greater heights above Earth's surface (thus taller turbines tend to be more efficient). The wind resource varies greatly by location. In some regions it is out of phase with energy demands—weak during the day but stronger at night; in other places, winds are stronger during the day. Transient weather patterns can bring hurricane-force gales or days and weeks of calm, when virtually no electricity can be generated.

Therefore when discussing solar panels and wind turbines it is important to understand the difference between *nameplate capacity* (how much power could be generated with constant sun or wind) and these resources' *average* power output (fig. 3.3). The ratio of these numbers is the *capacity factor.* A coal- or gas-fired baseload power plant might have a capacity factor of 90 percent; wind farms have capacity factors ranging between 22 and 43 percent.[8] In the United States, PV systems have capacity factors ranging between 12 and 20 percent, depending on the location.[9]

Uncontrollable resource variability is a problem for grid operators who need to match electricity generation with demand on a minute-by-minute basis. Daily and seasonal demand cycles are fairly easy to predict in general terms: electricity use tends to peak in the afternoons and dip at night; and in most temperate and tropical regions it increases during the hottest part of the summer when air conditioners are in use. Solar output tends to follow this cycle fairly well up to a point, but often cannot be dispatched to meet a surge of demand or turned off if demand is low (more recently built PV farms have "spinning" reserves where some proportion of the power output must be available for ramping). Wind power's variations often balance out those of solar; but sometimes both reinforce one another, producing an unusable surge of electricity that grid

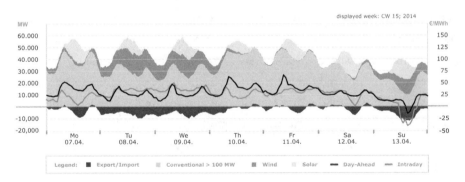

Figure 3.3. Intermittency of renewable energy electricity generation and its effect on price. This chart shows Germany's electricity production and spot prices for the week of April 7, 2014. As renewable energy production fluctuates, conventional production and the spot prices respond.
Source: Johannes Mayer, "Electricity Production and Spot Prices in Germany 2014" (Freiburg: Fraunhofer ISE, December 31, 2014), https://www.ise.fraunhofer.de/en/renewable -energy-data.

operators must somehow shed. And sometimes both sources are in a lull (the weather is cloudy and still), even though electricity demand is high. (Modern wind farms also have grid benefits, since they can be damped easily, which is useful for reactive power [voltage] control.)

Intermittency has long been recognized as a hindrance to the adoption of solar and wind technologies, and so a lot of thought has gone into finding ways to reduce or buffer that intermittency. Also, many countries now have experience integrating solar and wind into their grid systems. In short, there are strategies for dealing with intermittency—though each has limitations and costs.

Storage

The most obvious way to make up for the variability of solar and wind energy is by storing energy when it is available in surplus so that it can be used later. There are several ways energy can be stored, but before we survey them it will be helpful to know a little about how to evaluate storage systems.

Let's start with two factors: (1) the amount of *energy* the system can store (as expressed in watt-hours), and (2) the amount of *power* the system can absorb or deliver at any moment (as expressed in watts). A system that stores lots of energy won't be very useful if it can only receive or return that energy a little at a time. And a system with enormous power won't be helpful if it needs recharging after only a few minutes. Storage systems need to do well in both respects.

Energy density is especially relevant for alternative ways to power transportation. For electric vehicle (EV) batteries, it is useful to know the energy density both by weight (megajoules per kilogram, MJ/kg) and by volume (megajoules per liter, MJ/L). EVs are often burdened by heavy batteries (the battery pack of a Tesla Model S, for example, weighs in at over 1300 pounds). On the other hand, storing energy in the form of compressed hydrogen takes up a lot of space (see fig. 1.3).

Another metric of energy storage has to do with economic and environmental factors. What's the carbon footprint of a given storage technology? How much energy was used to construct it? And what's the energy cost of maintaining the technology over its projected lifetime? These three questions are closely related. Researchers Barnhart and Benson at Stanford University have proposed using the metric *energy stored on investment* (ESOI) as a way of tackling these issues.[10] It expresses the amount of energy that can be stored over the lifetime of a technology, divided by the amount of energy required to build that technology. The higher the ESOI value, the better the storage technology from an energy point of view—and, most likely, from an environmental perspective as well.

A final consideration with regard to energy storage has to do with limiting resources, such as lithium for batteries. For electricity, the three most widely discussed options for energy storage are geologic storage, hydrogen, and batteries.

Geologic Storage: Water Reservoirs, Compressed Air in Caverns

In the most common instance, this means pumping water uphill into a reservoir when electricity is overabundant, then letting it run back downhill to turn a turbine when more electricity is needed. Pumped storage is the most widely used grid-scale energy storage option; yet, for the United States, current pumped storage capacity is roughly 2 percent of the capacity of the electric grid.[11]

Pumped hydro power station. (Credit: A. Aleksandravicius, via Shutterstock.)

Pumped storage is the cheapest option for grid-scale energy storage (batteries have much higher embodied-energy costs). Barnhart and Benson determined that a typical pumped hydro facility has an ESOI value of 210,[12] which means it can store and deliver 210 times more energy over its lifetime than the amount of energy required to build it. Storage of compressed air in underground caverns also has a high ESOI value; however, this option is today rarely used.

The limits and downsides to geologic storage include the fact that it works only for stationary systems (not vehicles). It also suffers from low energy density: physicist Tom Murphy points out that "to match the energy contained in a gallon of gasoline, we would have to lift 13 tons of water (3500 gallons) one kilometer high (3280 feet)."[13] Therefore we would need a lot of reservoir volume to store really significant amounts of energy. But geologic storage requires appropriate topographic and geological conditions. In the final analysis, it is unclear whether it can be expanded enough to store anywhere near the amounts of energy we might need in an all-renewable future.

Hydrogen

Using electricity to produce hydrogen, then storing the hydrogen, offers another possible vector for buffering out the intermittency of renewable energy sources. Current hydrogen storage is minuscule. However, some analysts suggest hydrogen storage could be used widely at the household scale to store a large total amount of energy that could be flexibly used.[14]

Pellow et al. have determined that a hydrogen energy storage system would have an ESOI rating of 59,[15] which is much lower than the figure for pumped storage but higher than that of the best battery technology available today. Nevertheless, Pellow et al. also found that the low round-trip efficiency of a regenerative hydrogen fuel cell (RHFC) energy storage system "results in very high energy costs during operation, and a much lower overall energy efficiency than lithium ion batteries (0.30 for RHFC, *vs.* 0.83 for lithium ion batteries)."[16] Hydrogen storage represents a relatively efficient use of *manufacturing energy* to provide storage. But its *operational efficiency* must improve before it can compete with batteries in that regard.

In sum, hydrogen may be economic in some applications. It is potentially ...er than batteries for large-scale storage, and it can be adapted for use in vehicles and homes—though operational energy losses remain a problem.

Batteries

There is much ongoing research into the technology of converting electrical energy for storage as chemical energy in a battery. Just a couple of decades ago, lead-acid batteries (invented in 1859) were the primary available option for large-scale applications; today nickel- and lithium-based batteries are also available. Batteries are getting cheaper and better. In 2015, Tesla Motors Inc. unveiled a new generation of patented lithium-ion batteries designed for home and industrial use to store energy from sun and wind. This provoked speculation that higher volume production and further technical improvements could yield batteries cheap and powerful enough to solve the intermittency problems of renewable energy.

Since battery costs and efficiencies are a moving target, perhaps it is useful to consider the physical limits to battery improvements. Science writer Alice Friedemann has performed the thought experiment of examining the periodic table of elements to identify the lightest elements with multiple oxidation states that form compounds (oxidation-reduction reactions generate a voltage, which is the basis of electric cells or batteries). Ignoring problems such as materials scarcity, she finds that the theoretical upper energy density limit to the best materials would be around 5 megajoules (MJ) per kilogram (kg).[17] The best batteries currently commercially available are able to achieve about 0.5 MJ/kg, or 10 percent of this physical upper bound. Improvements would also be required in materials such as electrolytes, separators, current collectors, and packaging. Given all this, Friedemann concludes that "we're unlikely to improve the energy density by more than about a factor of two within about 20 years." Energy density is primarily a limiting factor in batteries for mobile purposes; still, for stationary purposes, low energy density implies the need for more material, and therefore typically translates to greater energetic cost in manufacturing.

The ESOI of batteries is quite low compared to that of pumped storage, and lower than that of hydrogen. Lithium-ion batteries perform best, with an ESOI value of 10.[18] Lead-acid batteries have an ESOI value of 2,[19] the lowest in the Barnhart and Benson study.

Batteries imply an added energy cost; what happens when this energy cost is added to the energy cost of building and installing renewable energy generation systems? Clearly, it reduces the energy "affordability" of the system; but if you're starting with an energy source that has a high EROEI, this is less of a problem. Using EROEI analysis, Charles Barnhart et al. found that storage is less "affordable" for PV than it is for wind.[20] Also, the manufacturing of batteries adds to carbon emissions. Technology writer Kris De Decker performed a life cycle analysis on existing PV-plus-batteries generating systems and found that they entail lower carbon emissions than conventional grid power, but not that much less.[21]

For small rolling vehicles and off-grid, self-contained electricity systems, batteries may provide the best available energy storage solution. Nevertheless, low energy density and low ESOI appear to be inherent drawbacks for chemical storage of electricity on a large scale; and while improvements are on the way, they are unlikely to change the overall situation.

Other Storage Options

While geologic storage, hydrogen, and batteries are the options most often discussed, there are others, such as compressed air canisters (for cars) and flywheels (for the grid); however, these are not widely used and are unlikely to offer substantial improvements over our three main candidates.[22]

There has also been talk of storing energy in electric fields (by way of capacitors) or magnetic fields (using superconductors). A company called EEstor claims a new capacitor capable of storage of 1 MJ/kg, which is about twice as good as the best current battery. Electromagnets of high-temperature superconductors can theoretically achieve about 4 MJ/kg. The ultimate physical potentials for such storage technologies would represent improvements over existing batteries but would still lack the energy density of hydrocarbon fuels.

Electrical energy could also be stored in synthetic fuels more chemically complex than hydrogen, including liquid fuels. These would offer greater energy density than battery storage and would therefore be better suited for use in vehicles; however, they would suffer from energy conversion inefficiencies.

In a recent paper, Mark Jacobson et al. propose the use of yet another storage medium—underground thermal energy storage (UTES).[23] Industrial waste heat, or heat from combined heat and power (CHP) plants or solar thermal collectors, would be channeled to storage tanks of water, pits of water, or fields of boreholes up to 300 m deep. For solar thermal plants, heat would be collected in the summer and released and used in winter. The heat is primarily used for space conditioning, though it can also be used for power generation, depending on the storage temperature. The technology (which is currently in use on the fifty-two-household Drake Landing Solar Community in Alberta, Canada, with 25,000 square feet of solar collectors) has high investment costs (3,400–4,500 euros/kW) but fairly low operation and maintenance costs.

Scaling up this technology is likely to be a big challenge. UTES (or any thermal energy storage design) is best used and optimized when done in conjunction with new construction or renovations; but given that the average building lifetime in the United States is 75 years, the rate of penetration growth is likely to be inhibited. A joint technical paper on the subject by the International Energy Agency (IEA) and the International Renewable Energy Agency (IRENA) confirms this is the case for Europe, where building stock turnover is only 1.3 percent per year and the renovation rate is only 1.5 percent per year.[24] It would be very expensive to try to retrofit existing buildings to take advantage of this process. It is also unclear how it could be fit into an existing dense urban area. UTES design is site specific, and subsurface storage technologies are site specific. This adds to cost and complexity.

UTES is also characterized by low energy density. Water-based systems can achieve up to 50 kWh/m^3 (180 MJ/m^3 or 0.18 MJ/kg) which is about at the level of a Li+ battery. The consequence of that is low area density (a scheme in Crailsheim, Germany, for 260 houses, one school, and one sports hall uses 79,000 square feet of solar collectors, 3500 cubic feet of peak load storage, 17,000 cubic feet of buffer storage, and 1.5 million cubic feet of borehole storage with 80 probes), and of course entails a lot of drilling, along with large quantities of probes, pipes, and other equipment. The IEA/IRENA technical paper notes that the barriers include system integration, regulation, high costs, material stability, and complexity, while R&D is needed for insulation and high-temperature materials.

Currently only 8–10 gigawatts (GW) of sensible thermal energy storage exists in the world, but Jacobson et al. propose capacity sufficient to support 467 terawatts (TW) of charge from solar thermal collectors. To say that this is a highly ambitious proposal may be an understatement.

The bottom line for energy storage: many options exist, and research is likely to expand their number and improve them. But each of the categories of options is subject to limits and costs, even assuming substantial technical improvements. Given different criteria (energy density, carbon emissions, cost), some storage options offer advantages over others. However, current electricity storage is only a tiny percentage of the amount that will likely be required in an all-renewable energy future—we need to build *a lot* of storage. And supplying large amounts of storage will add significantly to the financial, materials, energy, and carbon costs of systems.[25] A real-world example: California's Energy Storage law AB2514 directs utilities to install 1.3 GW of storage capacity by 2020. Total installed generation capacity today is 78 GW, of which 12.26 is renewable (excluding large hydro). The law says storage must be economically feasible, but utilities have so far balked at implementing it.

Grid Redesign

The electricity grids of the twentieth century were designed to distribute power from large, centralized coal, gas, nuclear, and hydro generating plants to far-flung end users. Grid managers learned to track electricity demand patterns (usually based on times of heavy use of domestic heating and air-conditioning), which tend to feature daily peaks. These demand spikes are now met by *peaking power generators* (usually fired by natural gas) that are used only for short periods each day. The low utilization of peaking generators, along with the necessary redundancy in the electricity grid, results in high costs to the electricity companies, which are passed on to customers.

The renewable electricity system of the twenty-first century will be different: it will accommodate numerous smaller and more geographically distributed power inputs, most of which are uncontrollably variable. Meeting demand will require, among other things, significant smart grid upgrades. The term "smart grid" doesn't refer to a specific technology, but rather to a set of related technologies whose goals are to gain a better understanding of what is happening on the grid in order to reduce power consumption during peak hours and incorporate grid energy storage, both of which make it easier to integrate more solar and wind. Disregarding the renewable energy transition, smart grids are expected to deliver increased efficiency and reliability, saving grid operators and consumers money. Add distributed renewable power generation, and the grid may evolve beyond a centralized system to become something of a collaborative network of electricity producers and consumers.

The main elements of a smart grid consist of integrated communications, sensing and measurement devices (smart meters and high-speed sensors deployed throughout the transmission network), devices to signal the current state of the grid, and better management and forecasting software; as renewable energy inputs are added, energy storage systems will inevitably become part of the network. Smart grids with a large share of renewables will also need additional transmission capacity to move more power longer distances to balance loads as output from distributed solar and wind generators varies.

A paper from Siemens Corporate Technology in Germany weighs the relative contributions of grid extensions and electricity storage to a hypothetical 100 percent renewable European grid, and finds that, with storage, renewables could supply up to 60 percent of power without additional grid capacity or backup, and 80 percent with an "ideal" grid.[26] These conclusions are similar to those of a National Renewable Energy Laboratory (NREL) study, which relies heavily on dispatchable biomass power generation to achieve the renewable target (about 15 percent biomass generation in 2050). They note regarding the grid that "electricity supply and demand can be balanced in every hour of the year in each region with nearly 80% of electricity from renewable resources, including nearly 50% from variable renewable generation, according to simulations of 2050 power system operations."[27]

How much will all this cost? A 2011 study by the Electric Power Resear Institute (EPRI) found that smart grid upgrades in the United States would require the investment of between $338 billion and $476 billion over the next 20 years, but would deliver $1.3 trillion to $2 trillion in benefits during that period.[28] Another study, this one by the U.S. Department of Energy, calculated that a more modest modernization of U.S. grids would save between $46 and $117 billion over the same twenty-year timeframe.[29]

Assuming that smart grid investments are a good deal over the long run, who pays for these upgrades over the short term? Experts disagree on whether recovery of a utility's smart grid upgrade costs should come from raising rates to customers or from some "nontraditional" source, such as government. There is also concern that utilities and regulators are accustomed to buying power equipment that lasts 40 years or more, whereas some electronic sensors and communications devices now being installed on the grid may last half that time, or as little as a decade.[30]

The electricity grid has been described as the largest machine ever created by human beings; as we make it larger and smarter in order to accommodate more variable and distributed renewable energy inputs, we also make it even more complex. Is there another solution? There is: do away with the centralized grid altogether and have energy generation and storage happen at the scale of communities. This would require every city and possibly every neighborhood to have enough generating and storage capacity, as well as needed control equipment, to sustain itself. The result would likely be a more expensive electricity system overall, and one that would, left entirely to the free market, result in much greater energy inequality (a subject to which we will return in chapter 8), since some households and communities would be able to afford robust systems, others none at all. The intermittency of wind and sunlight would also likely pose a greater challenge for more localized minigrids, unless they were linked over large geographical areas to take advantage of distant resources to make up for local shortfalls.

Decentralizing the grid would encourage energy use more in line with natural flows of renewable energy; also, households/communities would be more self-sufficient, and the system would entail less complexity and fewer interdependencies, resulting in less vulnerability to breaks in a brittle system. In light of all the factors mentioned, the likely outcome will be some mix of both centralized and decentralized grid systems, combining long-distance transmission infrastructure (high-voltage lines) with local distribution.

Demand Management

Given electricity sources whose unpredictably variable output doesn't coincide with times when electricity is typically used, one set of solutions (which we have just discussed) aims to make that output more predictable using *storage and control systems*; another set of solutions, generally referred to as *demand response*, is geared to manage *when* consumers use energy and *how much* they use, through voluntary programs or economic incentives. Although the purpose of demand response programs today is to avoid construction of costly generation capacity to meet peak demand, the practices are similarly applicable to managing the increased penetration of variable electricity generation such as solar and wind. Aligning electricity demand with supply entails two main substrategies: dynamic pricing, and smart appliances and equipment. One potentially important example of the latter is the use of electric car batteries for grid storage, discussed later in the chapter.

Dynamic pricing—changing the price of electricity according to its hour-by-hour availability—has led large industrial and commercial users to shift their usage to times when supplies are abundant and prices are low. This requires knowing when those times are, which in turn requires ways to communicate with users. In California, links between the independent system operator (ISO)—which coordinates, controls, and monitors the operation of the electrical power system—and large interruptible users have already been established; further, one of the goals of smart meter programs is to communicate real-time pricing information to residential and commercial customers so they can shift usage times. As this requires sensors, communication links, software, and data management, dynamic pricing is inseparably connected with the project of redesigning the grid, discussed earlier in this chapter. Using dynamic pric-

Battery of a Toyota 86 electric vehicle. (Credit: Tokumeigakarinoaoshima, via Wikimedia Commons.)

ing to enlist market forces in demand management will unquestionably help reduce times of over- or undersupply of electricity, thus increasing power affordability.[31]

Dynamic pricing can happen with the old grid infrastructure, it just requires feedback of the spot price to consumers who face that price. However, most domestic consumers currently don't have an easily accessible way to track the spot price in real time, or are subject to flat-rate pricing, and thus have no incentive to change their usage patterns.

When we're at home, we don't check electricity prices hour by hour to see when prices are high or low. How, then, can residential electricity customers be integrated into dynamic pricing programs? By automating the process via the so-called Internet of Things. Once most appliances are computerized and connected by wifi or hard line, they could in principle be set to respond to data from the utility company so they adjust their energy usage based on the current price of electricity. (Another potential for grid demand management entails allowing the utility to dial down power usage of appliances like refrigerators and air conditioners remotely during peak times.) This doesn't require a smart meter; in fact, most smart meters don't have this capacity. All that is required is a switch that the utility can turn on and off.

There are, of course, limits to these strategies. The Internet of Things implies additional material resources—which require extraction, manufacturing, transport, and operation—and also increased system complexity. It also raises privacy issues: already televisions are tracking (and potentially selling) users' usage data. Finally, some electricity usage is easily amenable to demand shifting; at home, for example, we may be quite willing to load up the washing machine, set its dial, and wait for the machine itself to determine when to wash our clothes based on hourly electricity price fluctuations. But if we're working at a computer, we might be less than pleased to see its screen go black following the momentary display of a message reading, "Sorry, electricity prices have just gone up."

Among smart appliances, electric cars have often been touted as having the greatest potential for helping match grid electricity demand with supply. Since automobiles are parked an average of 95 percent of the time, if EVs were left plugged in during that time electricity could flow to power lines and back, with a value to the utilities of up to $4000 per year per car.[32] The use of EV batteries to provide decentralized storage of electrical energy, either by delivering electricity into the grid or by throttling their charging rate, is known as vehicle-to-grid (V2G). Grid managers could incentivize vehicle owners to participate in V2G programs by offering discounted electricity at night to charge vehicles, and by offering fees to offset the cost of battery wear and tear from additional cycling. It is unclear, however, whether such incentives could realistically be greater than the value of the batteries to their owners.

Since proposed V2G programs center on the use of batteries for storage, all of the limits to battery storage technology previously discussed apply here. Currently, only pilot V2G programs exist, and the number of EVs in use worldwide is still too small to provide much real-world data on the likely benefits and drawbacks of a program large enough to impact grid reliability and price stability.

Capacity Redundancy

Another way to reduce the impact of energy source intermittency is to add redundant generation capacity: when the sun isn't shining and the wind isn't blowing, then simply rely on other electricity sources, which can be throttled down when sun and wind are abundantly available (this is already done with natural gas generators, though using them this way is much less energy efficient than as combined cycle *base load*, in which they operate continuously and are available 24 hours a day). Redundancy obviously adds to total system costs, and therefore proposals for future 100 percent renewable electricity systems typically attempt to reduce the need for it with strategies already discussed (storage, grid upgrades, and demand management). Nevertheless, capacity redundancy is the primary strategy that currently enables intermittent renewables to be integrated into electricity grid systems.

So far, solar and wind have remained proportionally small contributors to overall electrical energy in most nations, and variability has been buffered primarily by fossil energy resources (especially by natural gas–fired peaking plants, which can be powered up or down quite quickly). In effect, the grid itself becomes the battery for solar and wind generators. Renewable energy resources other than solar and wind could fill more of that role; these would likely include biomass, hydro, and geothermal. But are these resources up to the job? Let's take a look at each in turn.

Biomass

Burning wood, crop residues, and biogas is a dispatchable electricity source: as with coal or natural gas, if more electricity is needed then it's just a matter of firing up the boiler and adding fuel. However, this resource is limited, and long-term sustainability is uncertain. Forests cover 7 percent of Earth's surface, but net deforestation is occurring around the globe, especially in South America, Indonesia, and Africa.[33] The use of ever-larger areas of land and quantities of water for growth of dedicated "energy forests" also raises concerns about competition with food and fiber crops.

World electric power generation from biomass was about 405 TWh in 2013 from an installed capacity of 88 GW, with much of the growth based on a growing international trade in wood pellets (at some distance from the source, transport of wood pellets consumes more energy than the pellets will deliver).[34] Cogeneration or CHP plants can burn fossil fuels or biomass to generate electricity while also using their "waste" heat for space or water heating (biomass CHP is more efficient at producing heat than electricity, but it can be practical if there is a local source of excess biomass and a community or industrial demand nearby for heat and electricity). Most biomass generation plants are located in northern Europe, the United States, and Brazil, with increasing amounts in China, India, and Japan, and capacity has been growing at over 10 percent per year over the last decade.[35] However, biomass power plants are only half as efficient as natural gas plants, and they are limited in size by a fuel-shed of around 100 miles. Except in cases of long-distance trade in wood pellets, biomass availability is highly seasonal, and biomass storage is particularly inefficient with high rates of loss due to degradation.

In its favor, biomass is well suited for use in small-scale, region-appropriate applications where using local biomass is sustainable. In Europe there has been steady growth in biomass CHP plants in which scrap materials from wood processing or agriculture are burned, while in developing countries CHPs are often run on coconut or rice husks. Burning biomass and biogas is considered to be carbon neutral, since, unlike fossil fuels, these operate within the biospheric carbon cycle, though the increased reliance on wood as fuel raises concern about the time lag between combustion of the wood and the pace of carbon reuptake in new growth.

While biomass is a renewable resource, it is not a particularly expandable one. Often, available biomass is a waste product of other human activities, such as crop residues from agriculture, wood chips, sawdust and "black liquor" from wood products industries, and solid waste from municipal trash and sewage. In a less-fossil-fuel-intensive agricultural system, such as may be required globally in the future, crop residues may be needed to replenish soil fertility and won't be available for power generation. There may also be more competition for waste products in the future, as manufacturing from recycled materials increases.

Hydroelectric Power

Hydro dams have the potential to produce a moderate amount of additional, high-quality electricity in less-industrialized countries, but they are often associated with severe environmental and social costs. Particularly in northern Europe, hydropower already serves to balance the growing proportion of variable renewable electricity production, though hydropower itself can be subject to strong seasonal variations, which may be exacerbated by climate-change-induced changes in rainfall. Globally, there are many undeveloped dam sites with hydropower potential, though there are far fewer in the United States, where most of the best sites have already been dammed. With over 1000 GW of hydropower capacity installed globally,[36] the International Hydropower Association estimates that about one-third of the technical potential of world hydropower has already been developed.[37]

Geothermal

Geothermal energy is derived from the heat within Earth. It can be "mined" by extracting hot water or steam, either to run a turbine for electricity generation or for direct use of the heat itself. High-quality geothermal energy is typically available only in regions where tectonic plates meet, where volcanic and seismic activity are common, and where heat is fairly close to the surface. Currently, the only places being exploited for geothermal electrical power are ones where hydrothermal resources exist in the form of hot water or steam reservoirs. In these locations, hot groundwater is pumped to the surface from wells 2–3 km deep and is used to drive turbines. In theory, power can also be generated from hot dry rocks by pumping turbine fluid into them through boreholes that are 3–10 km deep. This method, called enhanced geothermal system (EGS) generation, is the subject of ongoing research and the construction of demonstration plants, and the first grid-connected commercial plant of 1.7 MW capacity came online in Nevada in 2013 as part of an existing geothermal field.[38] Because EGSs use fluid injection to open existing rock joints, there is some concern that this technology could generate earthquakes as an unintended side effect.[39] In general, early high hopes for EGS technology appear not to be panning out.

In 2013, world geothermal power capacity reached 12 GW with output rising to 76 TWh.[40] Annual growth of geothermal power capacity worldwide has slowed from 9 percent in 1997 to 4 percent in 2013.[41] Geothermal power plants produce much lower levels of emissions and use less land area than fossil fuel plants. However, technological improvements are necessary for the industry to continue to grow. Water can also be a limiting factor, since both hydrothermal and dry rock systems consume water.

There is no consensus on potential resource base estimates for geothermal power generation. Hydrothermal areas that have both heat and water are rare, so the large-scale expansion of geothermal power depends on whether lower-temperature hydrothermal resources can be tapped. A 2006 Massachusetts Institute of Technology (MIT) report estimated U.S. hydrothermal resources at 2400 to 9600 EJ, while dry-heat geothermal resources were estimated to be as much as 13 million EJ (as you'll recall, the world currently uses over 500 EJ per year), but the U.S. Department of Energy estimated in 2014 that technical advances needed to access the latter may still be 10–15 years from commercial maturity,[42] which may reflect inherent problems with EGS.

Biomass, hydro, and geothermal are probably the best three renewable electricity sources available for base load renewable power, though there are others (notably tidal and wave generators, which currently produce only very small total amounts of electrical power). This book focuses on solar and wind as the main candidates for expansion of renewable energy because these are the sources with the most immediate capacity for growth. No doubt some combination of biomass, hydro, and geothermal, used as base load and/or backup capacity, can help buffer the intermittency of solar and wind, but since these sources (with the possible exception of geothermal) have limited prospects for expansion, this could ultimately also limit the total amount of energy production capacity in an all-renewable future energy regime.

How about buffering the intermittency of solar and wind with . . . more solar and wind? After all, even if the weather is cloudy and still in a given location, it might be windy and sunny a few hundred miles away. This kind of capacity redundancy would require more grid interconnections and, of course, more solar panels and wind turbines. Since we couldn't know far in advance which other regions would be likely to provide capacity redundancy for ours, we would need enough redundancy in several places, and enough transmission capability, to meet possible supply shortfalls. All of this adds to the system cost.

The actual experience of grid operators integrating solar and wind into the grid has led to an emerging consensus that the cost of integrating renewables will shoot up as solar and wind make up a very high percentage of grid power.[43] In the early stages of solar and wind build-out, it is fairly easy to incorporate new uncontrollable inputs into the grid because redundancy already exists in the form of coal, natural gas, nuclear, and hydro generation plants, which have plenty of capacity to balance out added variable renewable electricity and match it with demand—which is also variable. However, as total solar and wind input surpasses 30 percent of grid electricity, the costs of integration are likely to gradually increase. Past 60 to 80 percent, the need for storage and redundancy will likely explode. The goal of a near-100 percent renewable, grid-based electricity system is a subject of great controversy and research, but it remains theoretical, because no society has created one yet, except for a couple of small islands in the Canary Islands (El Hierro)[44] and Denmark (Samsø) where wind and pumped hydro have been deployed to serve their small populations; or Uruguay, which generates the great majority of its electricity from hydro power.

Scaling Challenges

If we're to achieve a 100 percent renewable electricity system soon enough to significantly mitigate climate change, we'll have to build fast. The good news is that solar and wind are already growing quickly, as we have already seen. The bad news is that there appear to be some financial, energy, and environmental hurdles in the path toward scaling up these sources at the rates needed.

The energy transition will be expensive. While some estimates suggest that a renewable energy regime will be more affordable than a business-as-usual pathway dominated by fossil fuels (especially so once the climate and health impacts of the latter are taken into account),[45] it is doubtful that the business-as-usual pathway is itself affordable. And health and environmental costs avoided do not translate to money in the bank ready to be invested in alternative energy projects. Estimates of the total cost of moving to an all-renewable global electricity system are too preliminary to be exact, but are nevertheless expressed in the tens of trillions of dollars.[46] Where will the money come from? If the utility industry simply replaces coal, natural gas, and nuclear plants as they reach retirement age with solar, wind, geothermal, hydro, and biomass capacity, then most of the capital cost of the transition would come from the utility industry using its usual financing methods. But, again, to achieve the speed of transition needed, we would also have to retire fossil-fueled plants that are still well within their projected operating lifetime. That would imply higher rates of investment than the utility industry is accustomed to. Also, as the need for storage, capacity redundancy, and grid expansion and redesign increase, these will impose still more added costs.

Until recently, rapid expansion of solar and wind has relied on incentives, including rebates to homeowners installing PV systems and feed-in tariffs (long-term contracts to buy electricity from renewable energy producers, typically based on the cost of generation rather than existing market prices for electrical power). But those incentives are being reduced, eliminated, or thrown in doubt in countries such as Italy, Spain, and the United Kingdom, and in states such as Kansas and Arizona. While the falling costs of wind and solar are making them more directly competitive with incumbent fossil electricity sources, the loss of government financial support would slow the renewables transition.

A recent MIT study of the prospects for solar electricity found that, due to factors related to intermittency, "Even if solar PV generation becomes cost-competitive at low levels of penetration, revenues per kW of installed capacity will decline as solar penetration increases until a breakeven point is reached, beyond which further investment in solar PV would be unprofitable."[47] Therefore further subsidies for renewables (or penalties for nonrenewables) would probably be required if this energy source is to be scalable to replace the bulk of fossil-fueled generation.

Financing for solar and wind generation is fundamentally different from that for coal and gas plants. In the former case, investment is almost entirely up-front; from then on the "fuel" is free and maintenance is relatively inexpensive. In the latter, the cost of building the generation plant is proportionally less, with ongoing fuel costs being factored into wholesale and retail electricity prices. There is an obvious advantage to solar and wind from an investment standpoint (no worries about fluctuations in fuel prices), but there is also a drawback: front-loading of investment means that the availability of low-interest credit plays a major role in making new wind and solar capacity affordable.

Incumbent coal and gas power plants have the advantage of a lower tax burden, as fuel costs can be deducted from taxable income, while solar and wind cannot benefit from this deduction. Property taxes can also be an issue for large solar and wind installations, which take up much more land per unit of generating capacity than fossil fuel plants.

Aside from these financial problems, there is also an *energy* hurdle to the rapid transition to renewable electricity. Just as the financial investment in solar and wind generators is front-loaded, so is the energy investment in their construction: from an energy perspective, these generators must "pay" for themselves over time. This means that, if lots of generation capacity is built too quickly, it may constitute an energy sink rather than a true net energy source until rates of installation begin to slow (fig. 3.4). A 2013 study by Benson and Dale at Stanford University, cited earlier, showed that solar PV generation capacity installed between the years 2000 and 2012 paid for itself in energy terms only toward the end of that period; this was due to the high rates of growth, the energy costs of panel production and installation, and the relatively low EROEI of PV.[48] Wind power, with its higher EROEI, is less subject to this problem; nevertheless, the principle still holds: if the rate of installation of an energy-generating technology whose energy costs occur almost entirely in the manufacturing stage is steep, then the net energy available from that technology during the ramp-up period will be significantly lower than the gross energy produced by the installed generators (fig. 3.5).

This also means that, from an energy standpoint, a rapid deployment of solar and wind generators will almost certainly be subsidized mostly by fossil fuels. Which in turn implies that, during at least part of the transition period, society will need *more* energy from fossil fuels than it is currently deriving—unless existing energy demand can be throttled down while a larger proportion of remaining fossil energy consumption is devoted to all the activities needed to build and deploy wind turbines and solar panels.

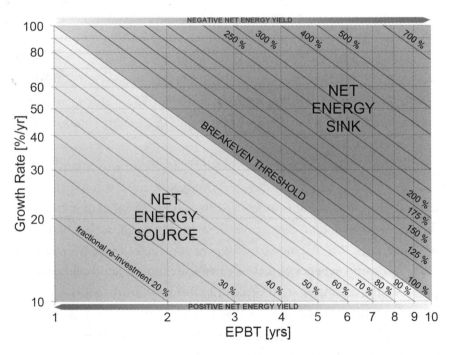

Figure 3.4. Conceptual energy balance. This figure shows growth rate [%/yr] as a function of energy payback time (EPBT) [yrs] for a number of fractional reinvestment rates [%] (diagonal lines).
Source: Michael Carbajales-Dale, "Fueling the Energy Transition: The Net Energy Perspective," GCEP Workshop on Net Energy Analysis at Stanford University, April 1, 2015.

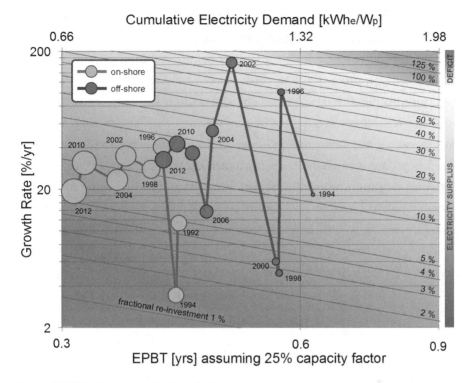

Figure 3.5. Wind energy payback period.
Source: Michael Carbajales-Dale, "Fueling the Energy Transition: The Net Energy Perspective," GCEP Workshop on Net Energy Analysis at Stanford University, April 1, 2015.

Another scaling challenge for solar and wind comes from the need for raw materials, including rare earth minerals for electromagnets in wind turbines and lithium for batteries. At current rates of installation this is not a significant barrier, but world supplies of these elements are limited and could constrain production; for example, at 10 percent annual growth in annual extraction rates, currently lithium reserves would last a mere 50 years.[49]

Questions about the technical potential of wind power pose yet another scaling challenge for renewable electricity sources. Early estimates of the potential ranged from ten to a hundred times current total world electricity generation capacity from all sources. Research by Adams and Keith notes "[w]ind resource estimates that ignore the effect of wind turbines in slowing large-scale winds may therefore substantially overestimate the wind power resource."[50] However, other researchers dispute this claim.[51]

Then there are location issues. Older-design wind turbines near urban areas have been reported to create low-frequency noises that are disturbing to at least some people (this is not a problem with offshore turbines—at least not for humans).[52] Solar panels can often be unobtrusively sited on rooftops, but producing really substantial amounts of energy from PV or concentrating solar will require real estate. Already, large concentrating solar thermal projects in the deserts of the American Southwest are forcing tradeoffs with habitat for species such as the desert tortoise. In addition, large solar arrays in desert areas require periodic washing of dust in order to maintain high levels of efficiency. Concentrating solar thermal plants need water for cooling as well, but this requires amounts of water that can be significant in these environments.[53]

Lessons from Spain and Germany

The world is still in the early phases of its renewable energy transition, but some clues about that transition's future, and some lessons on how to optimize it, can be gleaned from the experience of countries that have gone the farthest and fastest. Spain (with 27.4 percent of its electricity derived from solar and wind in 2014)[54] and Germany (with about 30 percent of electricity from renewable sources, including hydro)[55] are two of the leaders in this regard, but their stories are very different. And their efforts have both supporters, who characterize the transition so far as a great success, and detractors, who paint it as an expensive failure.

The Iberian Peninsula is sunny and has large wind resources; further, in terms of grid connections with other nations, Spain is relatively isolated. These factors together make the Spanish experience with PV and wind an interesting test case. Spain's experience with the rapid introduction of renewables started in 1997, with early strong support for solar and wind. The government instituted a standard offer (feed-in tariff) policy requiring that utilities purchase electricity generated by renewables at premium rates. Power companies, including Acciona, Endesa, and Iberdrola saw this an opportunity to start building their own wind farms. Spain's renewables subsidies led to a nearly fortyfold increase in wind capacity over the next dozen years, to 16.7 GW in 2008.[56]

In 2004, the Spanish government also instituted a generous feed-in tariff of 46 euro cents per kWh for solar. Again, investors rushed to cash in on this lucrative promise of long-term profits, and rates of solar installation soared. The government target for 2008 was 400 MW of new solar capacity; 3500 MW was actually installed. During 2008, Spain installed more than 2.5 GW of PV capacity, nearly half of the global total that year. At the same time, subsidies supported the construction of nearly 2 GW of generation capacity from large solar thermal electric plants.

Out of necessity, Spain pioneered the integration of large amounts of variable renewable electricity into the grid. The nation's grid operator, Red Eléctrica de España (REE), had argued that it would be impossible to integrate wind power at more than 12 percent of total electricity demand. However, in 2006 REE built a centralized dispatch system and required all wind farms to connect to it. This was the first system of its kind in the world, and it enabled Spain's wind power to grow to 20 percent of annual demand in 2014, providing over 60 percent of electricity at times of peak generation.[57]

The Solnova Solar Power Station near Seville, Spain. (Credit: Abengoa Solar, via Wikimedia Commons.)

But this rapid deployment of renewables meant the government was paying out more in subsidies than it had bargained for. An existing law that set limits on retail electricity rate increases required the government to make up for discrepancies between the utility industry's revenues and costs. By 2009 this rate freeze was causing Spain's utility system to run a deficit of 4 billion euros— roughly 20 percent above utility company revenues.[58] After the global economic crash of 2008, the Spanish government was simply unable to continue funding such deficits. The sitting center-left government reduced the feed-in tariff rates; in 2012 its center-right successors froze renewable energy incentives and introduced a complicated system that rewarded renewable energy producers even less.

Today Spain's renewable energy transition is moving very slowly. In retrospect, failures of the boom years can probably be chalked up to a combination of bad policies that failed to pay fairly for electricity and that lacked an upper limit on subsidies, and bad luck in the form of the global financial crisis.[59]

Germany offers a more encouraging example. Its *Energiewende*, or energy transition, has historical roots reaching back to the 1970s, when popular skepticism of nuclear power and support for renewables were already decisive political issues. Like their Spanish colleagues, German policy makers believed that early subsides for renewables would eventually lead to much lower prices for solar and wind—as they indeed have. But in Germany subsidies have been more consistently managed. Feed-in tariffs were instituted in 2004 and have been modified many times since. As of July 2014, subsidized rates for PV electricity ranged from 12.88 euro cents per kWh for small rooftop systems, to 8.92 euro cents per kWh for large utility-scaled systems.[60]

Today in Germany, wind, solar, and biomass combined account for almost the same portion of net electricity production as brown coal (biomass was 39 percent of the total).[61] Peak generation from combined wind and solar achieved 74 percent of total electricity production in April 2014.[62] In terms of generating capacity, Germany reached its 2010 target for wind power in 2005, its solar target for 2050 in 2012.

The 2011 Fukushima nuclear disaster in Japan led Germany's government to rethink the nation's reliance on nuclear power. Chancellor Angela Merkel

An Enercon wind farm in Lower Saxony, Germany. (Credit: Philip May, via Wikimedia Commons.)

announced the immediate, permanent shutdown of eight of its seventeen reactors and pledged to close the rest by the end of 2022. As a result, the largest four German utility companies—all owners of nuclear power plants—have seen declining electricity output. Meanwhile the nation doubled-down on its determination to develop renewable energy sources.

Germany has not only encouraged large-scale renewable energy systems but has also financed enormous numbers of distributed household- and community-sized generators. Six percent of German households were producing their own energy in 2014, and 20 percent said they aimed to do so by the end of the decade.[63] Compare this to California, where household solar ownership rates are about 1.2 percent.[64] Similarly, rather than relying only on grid-scale storage, Germany has created incentives for homeowners to add batteries to their residential PV systems.[65]

The *Energiewende* does have its detractors. A recent Wall Street Journal opinion piece noted, "Average electricity prices for companies have jumped 60% over the past five years because of costs passed along as part of government subsidies of renewable energy producers. Prices are now more than double those in the U.S. Yet nearly 75% of Germany's small- and medium-size industrial businesses say rising energy costs are a major risk, according to a recent survey by PricewaterhouseCoopers and the Federation of German Industry."[66] However, businesses are not fleeing the country as a result. In fact, it could be said that manufacturing is flourishing in Germany to a greater degree than in the United States, where electricity is so much cheaper: in 2012, industrial production made up 30.7 percent of the German economy, while it comprised only 20.6 percent of the U.S. economy.[67] Perhaps the biggest difference between critics and boosters of the energy transition is that critics assume that maintenance of the current largely fossil-fueled electricity system is a viable option, while boosters understand that, even with its challenges, the transition to an all-renewable energy economy is both necessary and inevitable.

What lessons can we take away from the examples of Spain and Germany? Subsidies for renewable electricity are still necessary, as are coordinated efforts to integrate and manage variable solar and wind inputs to the grid. These technical and economic issues are important, but perhaps less daunting than potential political roadblocks. As a recent analysis puts it, "The rapid deployment of large volumes of renewables requires both political will and a consistent policy."[68] When new governments overturn strong renewable energy policies instituted by previous governments, potential investors in wind and solar flee and may be shy to return. The nations that have had the most success with the renewable energy transition have implemented some form of feed-in tariff as a subsidy, and have stuck with that basic strategy even while adjusting tariffs somewhat as generation costs and other factors changed. Though solar and wind electricity prices have fallen significantly, it is difficult to imagine the renewables transition occurring at greater than old-plant replacement speed without such subsides or incentives.

Pushback against Wind and Solar

The recent rapid growth of wind and solar has posed problems for utility companies. As more and more solar and wind electricity generation capacity is installed—and this applies especially to rooftop solar—the utility companies' current business model faces an existential threat. Solar panel owners benefit from electricity free of generation costs, but utility companies have to pay for grid maintenance and are now forced to deal with uncontrollable energy inputs that may have to be offset, shed, or stored—and that costs money. The solar owner benefits, the utility pays.

Utilities are stuck with the bill for grid upgrades and grid-scale energy storage and, absent government subsidies, have no choice but to pass these costs on to customers in the form of higher rates. But then, facing higher grid rates, customers who can afford stand-alone solar systems may see it as being in their long-term advantage to go off grid. This hypothetical self-reinforcing feedback process has been called the utility death spiral.[69]

A 2010 study from the German Renewable Energies Agency concluded that nuclear power is inherently "incompatible with renewable energies."[70] Because solar and wind generators require no fuel, they can be the cheapest sources of electricity at the moment of production (their "levelized cost" includes payment on capital); therefore renewable electricity is often used as much as possible when it is available (though policies such as renewable portfolio standards [RPSs] play a role in this regard as well). When this happens, fossil-fueled and nuclear plants are throttled back if there is too much power relative to immediate demand—but not all power plants can do this. Older nuclear and coal power plants that can't be throttled back easily are therefore poorly suited for an electricity system with large and growing amounts of intermittent solar and wind power.

In the United States, utility companies—especially ones with large investments in nuclear and coal—have begun a coordinated campaign whose first phase included a push for state laws raising prices for solar customers. This has largely failed in legislatures around the country, as solar energy has proven popular even with political conservatives. More recently, the effort has centered on public utility commissions, where utility industry representatives have pushed for solar fee hikes, including high monthly charges for net metering, which pays solar customers for electricity they feed into the grid.[71]

Costs to utility companies from the introduction of distributed solar PV are somewhat balanced by the fact that added solar capacity helps reduce the strain on electric grids on summer days when demand soars and utilities must buy additional power at high rates. Nevertheless, as more residential and business customers install their own PV systems, revenues to the utility industry are starting to decline.[72] Industry-sponsored studies warn that the trend could eventually lead to a radical transformation of energy markets, on a scale similar to the restructuring of the telecommunications industry following the advent of the Internet and cell phones.

One partial solution is to entirely separate the businesses of power generation and grid operation (a situation that largely already exists in many places). That way, grid operators can concentrate on dealing with the task of optimizing the electricity system for renewable inputs, while nuclear, coal, gas, solar, and wind generators battle among themselves for market share. In any case, there is obviously a need for planning and policy at the governmental level to smooth the transition as much as possible.

∗   ∗   ∗

This rather lengthy chapter has explored issues surrounding the renewable energy transition in the electricity sector. It is in this sector where most of the growth in renewable energy has occurred so far. But we must not forget that only about 18 percent of final energy is consumed in the form of electricity globally (21 percent in the United States). As we have seen, even in this portion of the overall energy economy, substantial roadblocks to an all-renewable future remain (a very significant one that we will address later is the problem of embedded energy in the electricity sector—energy used in the processes of building and manufacturing solar panels, wind turbines, storage devices, and the rest of the infrastructure that will make up the renewable electricity system of the future). The next two chapters explore nonelectricity uses of energy, which pose their own, often greater, challenges.

CHAPTER 4

Transportation: The Substitution Challenge

L IQUID PETROLEUM is the world's dominant energy source. Oil is energy-dense, portable, and easily moved by pipeline and tanker—characteristics that have made it very well suited as a transportation fuel. Further, during the twentieth century it was amazingly cheap. During the 1980s, for example, a barrel of oil, which contains 1700 kWh of energy (the equivalent of over 10 years of hard human labor[1]), cost a mere $35 in inflation-adjusted dollars. Cheap transport energy helped fuel globalization, one of the most significant economic trends of the past few decades. Today, transportation accounts for 41 percent of U.S. energy end use, and over 95 percent of that transportation runs on oil.[2] The vast majority of cars still burn oil-derived fuels, as do airplanes, ships, trucks, and rail locomotives.

Trade depends upon the transport of raw materials and finished products. While movement of money can be effected electronically and almost instantaneously, the physical economy that money symbolizes requires wheels, roads, rails, rudders, landing strips—and the oil that lubricates and fuels transporta-

Combine harvesters in 1902 and 2014. (Credit, top: Robert N. Dennis, New York Public Library. Credit, bottom: (cc-by Martin Pettitt, via flickr.)

tion. Because oil has been so plentiful and cheap for the past century, we have globalized the production of most of the goods we depend upon. In 2011, U.S. ports took in $1.73 trillion in goods, 80 times the value of all U.S. trade 50 years ago.[3]

In addition, oil is critical to our industrial food system. While much food-system oil use is for the transport of farm inputs and outputs, diesel and gasoline are also used to power tractors and other on-farm machinery. Of the ten calories of industrial energy used to grow, transport, process, refrigerate, and cook the average calorie of food in the United States, 21 percent come from oil.[4]

If the transition to renewables is to succeed, it must address these systemic dependencies on liquid fuels. As we will see, there are efforts under way to do this, but enormous challenges remain.

Electrification

Since solar and wind energy technologies produce electricity, an obvious solution to our oil dependency for transport is to electrify transportation. There are currently roughly 750,000 road-legal electric vehicles (EVs) globally, and the rate of growth in the market is a spectacular 76 percent annually.[5] The United States has seen a growth rate of 69 percent annually, with about 300,000 vehicles now running on batteries.[6] At this growth rate (close to a doubling every year), the EV market in the United States could grow to equal the size of the current auto fleet in just a decade—though almost no one expects that to happen, as about half the gasoline-powered cars now in service will still be operational in ten years, and the vast majority of automobiles still being sold have conventional combustion engines.

Overhead and undercarriage recharging for electric buses. (Credit, top: Oliver.auge, via Wikimedia commons. Credit, bottom: Spsmiler, via Wikimedia commons.)

However, electric cars suffer from the inherent inefficiency of all personal motorized, road-based transport: the need to move a one- to two-ton vehicle in order to transport a few hundred pounds worth of people. The vehicles themselves represent large amounts of embodied energy, up to 75 MWh each.[7] That means today's global fleet of over one billion automobiles represents roughly 50 million GWh of energy, many times the total amount of renewable electricity produced in 2014.[8] Further, road building and maintenance also require energy: a meta-study at the University of Washington found that a reasonable estimate of total energy consumption for these purposes is between 0.5 and 1 GWh per lane-kilometer of paved roadway.[9] The vast majority of this energy expenditure is in the form of oil. With 65 million kilometers of paved roads worldwide, that represents up to 65 million GWh of energy, as much as is embodied in the world's cars.[10]

The strategy of electrifying transport entails both opportunities and obstacles. The opportunities start with the fact that 1 kWh of energy will propel a typical electric car 2.94 miles, compared to 0.83 miles for a similarly sized gasoline-powered vehicle.[11] That's because electric motors are very efficient compared to internal combustion engines.

However, the heart of the electric vehicle is its battery. And as we've already seen, though battery technologies are subject to innovation, falling prices, and increasing storage per unit of weight, nevertheless even the best theoretical battery has very low energy density compared to petroleum-based fuels. The result is that batteries work best in small, light vehicles. While a large majority of vehicles on the road are used to move people, 99.9 percent of the total weight being transported on U.S. roads (not counting the vehicles themselves) is goods that we consume.[12] But large, heavy vehicles such as trucks, tractors, and cargo ships require batteries too heavy to be practical in most instances, particularly if they are traveling long distances. Meanwhile, battery-powered aviation (except in the case of one- or two-passenger aircraft) is simply not an option.

Many railways are already electrified (in 2012 half of all rail tonnage worldwide was carried by electric traction), and this is accomplished without batteries: electricity is distributed to locomotives via an overhead line or a third rail. Electrified rail offers the advantages of greater efficiency and lower operating costs as compared to diesel power, though initial capital costs for electrification are relatively high. A few companies have proposed building electrified highways in order to bring similar efficiencies to electric trucking, though once again infrastructure costs would be significant.[13]

Electric city buses and streetcars that draw their power from overhead wires have been in use for over a century. A new generation of battery-powered electric buses (some of which recharge at bus stops) is gaining in popularity—including on some bus rapid transit (BRT) systems, which feature high-capacity buses operating on dedicated corridors, with off-board fare collection, and station platforms level with the bus floor. Electric long-haul buses and electric trains would work for intercity trips. Electric vans running as jitneys or taxis, electric car-shares,[14] and small electric buses would be ideal for interneighborhood trips.

Over the short run, transport electrification is likely to take the form of increasing numbers of electric cars, though that trend may ultimately be limited by the inherent inefficiencies of personal automobile ownership and operation, and by the need for oil in building and repairing highways. Electrified rail could theoretically replace trucking and domestic aviation and move many more people than it currently does; however, the United States—with its decrepit existing passenger rail networks—is at a distinct disadvantage in this regard. Substantially expanding electric rail and buses will entail up-front infrastructure costs, embodied energy, and time for build-out. This implies a multidecade effort involving large initial subsidies, presumably from government.

Biofuels

Where batteries are unsuited for transport (e.g., in heavy road vehicles, ships, and aircraft), why not use biofuels? These are renewable fuels made from food crops, agricultural and forestry residue, or algae, usually in the forms of ethanol, biodiesel, or methanol. The United States already produces 14.34 billion gallons of ethanol and 1.27 billion gallons of biodiesel annually,[15] roughly 10 percent of the gasoline consumed in 2014.[16] Could these fuels be used to run the nation's fleets of aircraft, ships, and trucks?

Since aviation, which represents 2 percent of global carbon emissions,[17] cannot be electrified, let's focus on this industry. The direct use of ethanol or biodiesel in current aircraft engines is impractical because these fuels do not have the right chemical characteristics (biodiesel, for example, tends to become highly viscous at low temperatures). However, the aviation industry is engaged in ongoing experimentation with several chemical pathways toward a replacement fuel, the most common of which is based on refining oils extracted from plants or from algae. Airlines were approved to use up to a 50 percent blend of such fuels in 2011, and both test and commercial flights have been successful. Research continues into an alcohol-based pathway, as well as through synthetic biology. Virgin Atlantic has announced its intention to test an alternative aviation fuel reputed to have half the carbon footprint of the standard jet kerosene from a production process that ferments carbon monoxide–rich gases from industrial steel production into ethanol, with further conversion to jet fuel using the LanzaTech process.[18]

The challenges for the growth of the biofuels industry are environmental tradeoffs, cost, scalability, and energy profitability. These challenges are all closely interrelated. Let's examine each of them, then return to a consideration of the task of running global aviation on a renewable replacement for refined petroleum.

Environmental Tradeoffs

Nearly all commercial biofuels are currently produced from food crops. In the United States, most ethanol is produced from corn, while Brazil grows sugar cane for this purpose. Global feedstocks for biodiesel include soybeans, palm oil, and jatropha. Producing these fuel crops has led to controversies about the diversion of land from growing food for people to making fuel for vehicles. Soil degradation, water use, and biodiversity loss are also implied in current agricultural biofuels production. Tellingly, in April 2015 the European Parliament voted to limit the use of crop-based biofuels due to impacts on food prices, hunger, forest destruction, land consumption, and climate change.[19]

Cost

New methods of biofuels production from agricultural and forestry residues, or from algae, have been proposed as solutions to the environmental tradeoffs of current biofuels production. However, so far these second- and third-generation biofuels have proven too costly to produce commercially. The U.S. Energy Independence and Security Act of 2007 established cellulosic (second-generation) biofuel mandates for succeeding years; each year, the Environmental Protection Agency has had to waive most of that mandate due to the inability of the industry to profitably produce sufficient fuel. Actual 2015 cellulosic biofuel production was less than 100 million gallons, compared to the original 3 billion gallon mandate.[20] Biodiesel from algae (a so-called third-generation biofuel) has likewise proven more difficult to produce profitably than was forecast by the industry, with the break-even price currently stuck at about $7.50 per gallon.[21]

Scalability

Unless and until second- and third-generation biofuels become commercially viable, the production of biofuels will continue to depend upon the conversion of land from forest or food production to fuel production. Given a growing human population with growing food requirements, as well as increasing concerns regarding loss of biodiversity, expansion of biofuels production beyond current levels creates unacceptable pressures for further conversion of land to dedicated fuel use.

In the case of jet fuel, the capacity to produce renewable alternatives based on plant-based oils is severely limited by the sheer magnitude of global jet fuel consumption. In 2012, global jet fuel use totaled about 250 million metric tons,[22] while worldwide production of all edible oils reached 161 million metric tons.[23] Given refining losses, diversion of even all global food oils to jet fuel production could supplant only about half of total current jet fuel use.

Energy Profitability

The last of these four factors heavily influences the previous two, and is generally acknowledged as the greatest economic hurdle to expanded use of biofuels. Calculated energy returned on energy invested (EROEI) figures for corn ethanol production in the United States range from less than 1:1 to 1.8:1—which falls below a proposed 3:1 threshold of economic viability for an energy resource.[24] Ethanol from sugar cane in Brazil is calculated to have an EROEI of 3.6:1 to 4:1, but when made from Louisiana sugar cane in the United States, where growing conditions are worse, the EROEI is closer to 1:1.[25] Distillation is highly energy intensive, and even more so in the case of cellulosic ethanol because the initial beer concentration is so low (~ 4 percent compared to 10–12 percent for corn). This dramatically increases the amount of energy needed to boil off the remaining water. At absolute minimum, 15,000 BTU of energy are required in distillation alone per gallon of ethanol produced (current corn ethanol plants use about 40,000 BTU per gallon).[26] This sets the limit on EROEI. If distillation were the only energy input in the process, and it could be accomplished at the thermodynamic minimum, then EROEI would be about 5:1. But there are other energy inputs to the process, and distillation is not at the thermodynamic minimum.

Soybean biodiesel currently returns 3.6 to 4 times the energy that is used to produce it, if co-products are credited.[27] Palm oil biodiesel appears to have the highest energy profitability, but it also has the highest environmental impacts. The EROEI of algal biodiesel has not been accurately calculated, since production is not yet occurring at an industrial scale; however, the high currently calculated break-even market price for the fuel suggests very low energy profitability.

When a great deal of energy has to be invested in an energy production process (in the case of ethanol, that investment includes plowing, seeding, fertilizing, harvesting, transporting, and distilling), this also implies greenhouse gas emissions. Thus, in the United States the use of ethanol to replace gasoline may reduce overall emissions only minimally.

Now, let's return to our discussion of the potential for biofuels in aviation. Clearly, it is physically possible to manufacture such fuels. However, doing so at the scale required to support the industry at its current size, without unacceptable environmental and social impacts, and at a sufficient energy profit so that fuels are affordable without massive financial and energy subsidies, will pose challenges at every step along the way.

Hydrogen

Hydrogen may be a practical fuel for transport uses in marginal applications. Toyota has unveiled the first mass-market hydrogen car, and California has already installed a tiny network of hydrogen fueling stations, which it promises to expand. Energy futurists have long predicted a "hydrogen economy," and some would say we are finally seeing the very first glimmer of its dawn.

Why has there been such a long wait? It turns out there are many potholes on the hydrogen highway. The first is the problem of getting the hydrogen:

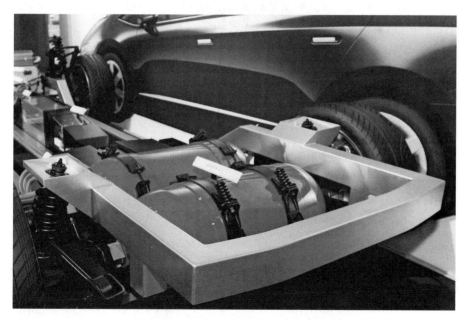

Hydrogen tanks in a Honda PCX prototype. (Credit: Morio, via Wikimedia commons.)

most commercial hydrogen is currently made from natural gas. Making hydrogen from water using renewable electricity implies substantial energy losses, with current hydrolysis systems averaging 65 percent efficiency.[28] Other production routes, such as biological water splitting, fermentation, solar thermal water splitting, and biohydrogen, are being researched, but none has achieved commercialization.[29]

The next hurdle is storage. Because hydrogen has a very low energy density per unit of volume, hydrogen-powered airplanes would need to carry compressed hydrogen in large storage containers that would add substantially to the size of the aircraft. Further, tanks will inevitably tend to leak, because the hydrogen atom is the smallest of all atoms and can eventually work its way through just about any material used to contain it.

Moreover, since conversion of energy is never 100 percent efficient, converting energy from electricity (e.g., from solar or wind) to hydrogen for storage before converting it back to electricity for final use will inevitably entail losses.

The problems with hydrogen are substantial enough that many analysts have concluded that its role in future energy systems will be limited (we are likely never to see a "hydrogen economy"), though for some applications it may indeed make sense. Energy storage using hydrogen fuel cells achieves a higher energy stored on investment (ESOI) ratio than battery storage (though not as high as pumped geological storage),[30] so some utility companies may end up using hydrogen storage systems for overgeneration in preference to battery banks.

Some 100 percent renewable energy plans are counting on cryogenic hydrogen to entirely solve the aviation problem, but the challenges of significantly replacing oil as a transport fuel with hydrogen are such that these should be viewed with caution. Not just fuel systems, but entire airplanes would need to be redesigned, with fuel tanks four times larger than today's, and the fuel would likely be significantly more expensive.

Hydrogen is also being considered as a fuel for ships.[31] In shipping, the requirement for larger fuel tanks might not pose as much of a problem as in aviation; but high fuel costs would perhaps be even more of a burden, to which the inevitable leakage of hydrogen during long voyages at sea could only add.

Natural Gas

The challenges of finding renewable substitutes for liquid fuels have led some analysts to consider the use of compressed natural gas (CNG) as a bridge fuel to a renewable future. It could have the advantage of producing somewhat less greenhouse gas emissions; and CNG could fuel long-haul trucks, earth-moving and mining equipment, buses, and tractors. Indeed, some conversions are already under way (thousands of natural gas–fueled buses are already on the road, and FedEx and UPS are currently refitting their truck fleets).

However, natural gas is of course a fossil fuel. That ensures the dual problems of continuing greenhouse gas emissions and depletion of the resource base. The latter may be decisive from an economic standpoint. Conventional wisdom holds that the United States has an abundance of natural gas as a result of the opening of shale reservoirs with hydrofracturing. However, nonshale natural gas production in the United States is in steady decline, and research at Post Carbon Institute suggests that a peak in domestic shale gas production is likely later this decade, followed by ongoing production declines.[32] A full repeat of the recent U.S. shale gas boom elsewhere in the world is unlikely due to a lack of infrastructure, expertise, and favorable private mineral rights ownership regimes. According to the independent German analytic organization Energy Watch Group, global natural gas production is likely to peak during the next decade.[33]

If renewable energy sources like solar and wind replace natural gas for power generation, this could free up some natural gas for the transport sector. However, even this reallocated resource availability would be temporary, perhaps providing a window of another decade in the United States—roughly the amount of time required to fully build out the new fleet of CNG vehicles. Thus, by the time the latter were ready to go in full force, their fuel supply would be uncertain. Natural gas could prove to be a bridge to nowhere.

Sails and Kites

Shipping consumes only 7.4 percent of oil used annually,[34] but it accounts for 90 percent of global trade, which in its current form would wither almost instantly in the absence of petroleum. Would it be possible to maintain high levels of trade using wind power, via sails or kites?

Kites can be flown at altitudes of 100–300 meters (330–980 feet), where winds are much stronger than at water surface; thus they receive a far higher thrust per unit area than conventional mast-mounted sails. SkySails, a Hamburg-based company, currently sells equipment to propel cargo ships, large yachts, and fishing vessels with sail-kites. Ships using the system maintain use of their oil-fueled motors, making them hybrid vehicles. The use of kite sails is estimated to reduce fuel consumption by 10 to 35 percent.[35] As use of the technology expands and evolves, and as ships are redesigned to use it more effectively, efficiencies may improve.

A more intensive return to wind-based ocean transport is being proposed by the Sail Transport Network,[36] which promotes both short-haul local deliveries using small boats in places such as the Puget Sound in Washington State, and major cargo deliveries using sail-powered or sail-assisted vessels. Meanwhile British wind power company B9 has tested the design elements of a planned 100-meter, 3000-ton carbon-neutral freighter that uses 60 percent wind power, relying on three computer-operated 55-meter masts supplemented by a biogas engine converting food waste into methane. B9 sees the plan as working best on small freighters.

A return to wind would almost certainly entail slower average speeds. However, in order to lower running costs during periods when oil prices are high, cargo ships are already accustomed to reducing speed to 12–15 knots, which makes them slower than the sail-powered clipper ships of the late nineteenth century.[37] Sailing boats also have to wait for the right winds, tides, and currents. In addition, they may need retractable masts if they have to go under bridges. Many of the seafaring skills of our ancestors may have to be rediscovered or relearned.[38]

Summary: A Less Mobile All-Renewable Future

The most likely adaptive strategy for the transport sector as we move toward an all-renewable future will entail all of the above. Farmers will likely use site-made biofuels to power agricultural machinery. Most cars will run on batteries, a few on fuel cells. CNG will be used for large vehicles until natural gas becomes too expensive or until engineers come up with a better option. Ships will employ more kites and sails. At least some aircraft will burn expensive, sophisticated biofuels, again until engineers find a better solution—if there is one.

This is not a very satisfying conclusion, for several reasons.

First, the electrification of transport (directly or via hydrogen or batteries) will put a significant extra burden on solar and wind technologies, requiring them to power not only much of the posttransition electricity sector but a large portion of the transport sector as well.

Second, we have not addressed the embodied energy in transport—the energy used in manufacturing cars and trucks; in building ships, locomotives, and aircraft; as well as in making roads, airports, rails, docks, and terminals. As problematic as it is to replace the operational energy for transport with renewable substitutes, the challenge for supplying manufacturing energy is probably even greater (as we will see in the next chapter).

Because oil is economically crucial and hard to replace, and because oil is leading the EROEI decline of fossil fuels, more and more energy investment capital will have to go toward maintaining essential existing oil-based energy usage systems, just as massive new investment is needed for renewable energy capacity, energy storage, and grid upgrades. The ballooning need for new investment just for *current* systems is confirmed (but probably seriously underestimated) in the 2014 International Energy Agency World Energy Investment Outlook report, which concludes that "meeting the world's growing need for [mostly fossil] energy will require more than $48 trillion in investment over the period to 2035."[39]

Thus, due to rising oil production costs and declining returns on investment, and the concomitant need to deploy often problematic or limited and expensive alternatives, society will probably become less mobile as the energy transition picks up speed. Even though big container ships use very little energy to move a ton of freight long distances, global trade of material goods will likely decline rather than inexorably grow. Nonproductive use of oil—the operation of personal vehicles and tourism for the middle class—will fare far worse. The implications of liquid fuel substitution limits for industrial agriculture are especially worrisome in a world of continually expanding human population. For transport, trade, and agriculture, renewable energy options exist—as we have seen—but they tend to be slower, more expensive, or supply constrained.

CHAPTER 5

Other Uses of Fossil Fuels: The Substitution Challenge Continues

T HIS CHAPTER EXPLORES three broad categories of energy use. The first considers ways in which energy becomes embodied in infrastructure and manufactured products (primarily through the use of high levels of heat). The second has to do with the creation of lower-temperature heat for heating buildings and domestic hot water. The third has to do with the use of fossil fuels as feedstocks for chemicals and plastics. As we will see, the second of these three (heat for buildings) is probably the easiest to address with efficiency measures and renewable energy, while the first (high-temperature heat for industrial processes) poses possibly the highest substitution hurdle of all for 100 percent renewable energy systems.

High-Temperature Heat for Industrial Processes

The production of many common materials—including steel and cement—requires extremely high heat (fig. 5.1). For example, making cement (the key ingredient of concrete) involves feeding crushed limestone, clay, sand, and other ingredients into a cement kiln kept at 1450°C (2700°F); kilns are often as much as 20 feet in diameter and up to 750 feet long.[1] Although the cement industry is responsible for only one-quarter of one percent of total U.S. energy consumption, it is the most energy-intensive of all manufacturing industries.[2] The main fuels consumed in the process are coal and petroleum coke, though natural gas and oil are also used. It's hard to imagine cement being made any other way, but it's also hard to imagine living without it: concrete is essential to nearly all building construction as well as to roads, dams, aqueducts—and pads for wind turbines.

The main ingredient in the making of steel is pig iron, which is in turn produced in a blast furnace. The primary fuel for the process is coke (made from coal). Steel is essential in the construction of transport vehicles, agricultural machinery, telecommunications infrastructure, and buildings—indeed, the entire scaffolding of industrial civilization. Steel is also used to make rebar, which is used in all forms of concrete to provide tensile strength. There is no modern concrete construction without steel. (Use of rebar also means that all concrete structures will inevitably succumb to corrosion.)

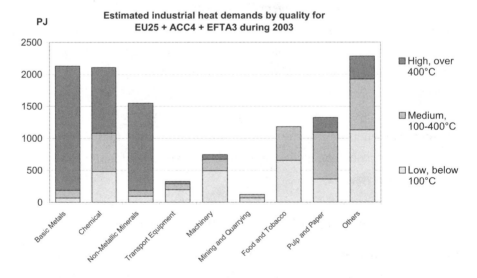

Figure 5.1. Temperatures used in industrial processes.
Source: Euroheat & Power, Ecoheatcool Work Package 1 (Brussels: Euroheat & Power, 2006).

The industrial processes that are used to manufacture renewable energy sources (wind turbines, photovoltaic panels, flat plate collectors, and solar concentrators) need high temperatures, as do factories that make electric trains, electric cars, computers, light-emitting diodes (LEDs), and batteries or their components. The production of glass uses temperatures up to 1575°C; the recycling of aluminum needs 660°C; the recycling of steel occurs at 1520°C; the production of aluminum from mined ores needs 2000°C; the firing of ceramics occurs at 1000°C to 1400°C; and the manufacturing of silicon microchips and solar cells uses heat at 1900°C.[3] Relatively little process heat currently comes from electricity, as figure 5.2 shows.

How could we obtain these high levels of heat without burning fossil fuels? There are four main options: electricity, solar thermal, burning biomass or biogas, and hydrogen. Let's consider each in turn.

Electricity

It is of course possible to create heat from electricity. We do so at the household level with electric heaters and electric cookers, and industrially with kilns and

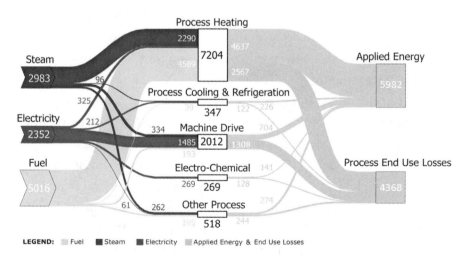

Figure 5.2. Sankey diagram of process energy flow in U.S. manufacturing sector.
Source: http://energy.gov/eere/amo/static-sankey-diagram-process-energy-us-manufacturing -sector.

arc furnaces that are used for producing cast iron and recycling steel. More-over, converting electricity to heat is efficient, in that nearly all the energy is converted.

But this way of using electricity tends to be *economically* inefficient. It takes 2 to 3 kWh of coal or gas thermal energy to make 1 kWh of electricity[4]; thus it is roughly two to three times cheaper to make heat by simply burning fossil fuels directly than to convert electricity to heat. This means if solar or wind power is at grid parity (i.e., if it costs about as much as power from a coal- or gas-fired plant), then using it to make heat will be two to three times more costly than burning coal or gas. As we have noted, electricity is high-quality energy, whereas heat is the lowest-quality energy: we currently use a lot of heat to make electricity. So to use electricity to make heat is a little like going to the trouble of turning gold into lead. Moreover, electrifying basic industrial processes while we are also electrifying a lot of transportation will add further to the already daunting task of producing all our electricity from renewable sources.

Solar Thermal

Many concentrating solar thermal generating facilities, which use arrays of mir-rors to focus sunlight on a small area, achieve very high temperatures. Parabolic trough systems only achieve temperatures of about 400°C,[5] but point concentra-tors (including parabolic dish systems, solar power towers, and solar furnaces) can produce temperatures up to 3500°C (6332°F), high enough to manufacture steel and cement, as well as microchips, solar cells, and carbon nanotubes. Fur-ther, this level of heat can be achieved in just seconds. The world's most power-ful solar furnace, in Odeillo, France, built in 1970, has a power output of 1 MW.

However, there are significant challenges to building industrial-scale solar thermal cement kilns and blast furnaces. It is unclear how a focal point of in-tense heat could be expanded to a controllable system. In a cement kiln, for example, the main calcining reaction takes place around 1000°C, whereas the major clinkering happens at 1400°C to 1500°C—all within the same 40-meter kiln. Perhaps molten salts could deliver heat of that magnitude, or perhaps ce-ment kilns could be redesigned to have a solar focal point at the current burn

Solar furnace in Odeillo, France. (Credit: Björn Appel, via Wikimedia commons.)

zone. For iron smelting, it's also unclear how the heat would be delivered to the furnace and maintained at the 2000°C level.

A perhaps greater challenge is the fact that these high temperature processes rarely shut down because cooling can badly damage the brickwork lining them. This raises the issue of heat storage to offset the problem of cloudy or rainy days or nighttime at our solar furnace; there would need to be very short-term storage to kick in if the sun goes behind clouds and delivered temperatures drop. But we don't have good ways to store large volumes of high-temperature heat (low-temperature heat is fairly easily stored in the form of hot water). "Storing work"—that is, just working when the sun is out—could be an option for some high-temperature processes but probably not for metal smelting and cement making; and it would make winter workdays very short at high latitudes.[6]

The problems of heat intermittency and heat storage suggest the efficacy of "batch" production over continual production. However, this would necessarily be production at lower volume, higher cost, and lower efficiency.

Finally there is the question of scale. The largest solar furnace in the world, as noted, is 1 MW, and it is a big installation with 60 heliostats. In the United States alone, heat demand in the residential, commercial, and industrial sectors combined is about 15 exajoules.[7] With a capacity factor of 20 percent, the solar furnace at Odeillo represents 1/4,000,000th of the heat demand in the United States; even if high-temperature heat is only one-third of the total heat demand, our world-leading solar furnace still provides just about one-millionth of the high-temperature heat used industrially in the United States. Providing the rest of that heat with solar furnaces would imply a lot of furnaces, with a huge footprint.

Biomass and Biogas

Prior to the introduction of coal, charcoal was widely used for smelting metals. In many respects it is superior to coal: charcoal burns hotter and contains far fewer impurities. Further, when it comes from a sustainable source, burning charcoal is carbon-neutral.

This raises the question of why the use of charcoal in metal smelting largely died out. We'll answer that in a moment. But first it is important to note that charcoal-based smelting still flourishes in Brazil, which has large iron deposits but little domestic coal. It is the world's largest producer of charcoal and the ninth biggest steel producer.

About half of Brazil's charcoal industry relies on plantations of fast-growing eucalyptus, cultivated specifically for the purpose, with the rest sourced from native forests through deforestation and from the use of sawmill by-products. While in medieval Europe charcoal-making was a cottage industry, Brazil has scaled up the process to encompass thousands of charcoal kilns operating at any one time.

But could other countries do what Brazil does? Probably not. During the nineteenth century, when charcoal was still widely used industrially in the United States and elsewhere, forests were being cut at a rate far above that of regrowth. Meanwhile, we were producing only a small fraction of the steel being made today. There is, quite simply, not enough forest in the world to enable this option to be deployed on a large scale. Just compare China's annual steel production (over 800 million tons) with Brazil's (34 million tons)[8] and consider the fact that Brazil's carbon emissions from steel production have increased in recent years due to deforestation, even though the proportion of coal used declined. To supply the charcoal needed by the steel industry entirely from renewable, plantation-grown trees, an additional 1.8 million hectares of land (4.4 million acres) would need to be dedicated to charcoal production.[9] And we haven't even considered using charcoal for production of cement and for other high-temperature processes.

Aside from charcoal, biogas could also theoretically do the job—if you could get enough of it. Methane can be harvested from landfills, or be produced in human waste, animal waste, or food waste digesters. The World Bioenergy Association estimates that biogas production could potentially grow to equal about one-third of the current global natural gas supply.[10] Remember, though, that we are also counting on biomass and biogas as electricity sources to balance the intermittency of solar and wind; and (in the case of biogas) possibly as a renewable transport fuel as well. Remember, too, that we are assuming reduced landfill waste in the renewable future and increased use of organic waste for agriculture.

Hydrogen

A recently published roadmap for 100 percent renewable energy[11] suggests the use of hydrogen for high-heat industrial processes. One discussion elsewhere in the literature with regard to cement production using hydrogen notes the following:

> Due to its explosive properties, hydrogen could not be used in existing cement kilns, but could principally be utilized after dilution with other gaseous fuels or inert gases like nitrogen or steam. Furthermore, the combustion and radiation properties of hydrogen differ significantly from those of the fuels being used today in the

cement industry meaning that—even if handling problems were solved—the clinker burning process would have to be significantly modified. By pre-combustion technologies, only CO_2 from fuel combustion, but not from limestone decarbonation can be captured.[12]

In essence, the substitution of fossil fuels with hydrogen in cement making (and other high-heat processes) is theoretically possible but would require a massive redesign of these processes. Estimates for the cost of cement production using hydrogen are inevitably vague and seldom include the costs of process redesign (plus testing, piloting, scaling, and commercialization) that would have to occur.

Using hydrogen for a wide range of industrial processes would require a scaling up of electricity production in order to make the hydrogen, again adding to the already significant challenge of replacing existing fuels in the electricity sector. In addition, there would be costs and challenges associated with storing hydrogen (which, as mentioned earlier, takes up significant space and has a pronounced tendency to leak). Also, because it is so "leaky," every industrial plant using it would have to build and operate its own hydrogen production plant as well, since hydrogen can't be centrally produced and distributed by pipeline like natural gas.

Low-Temperature Heat

Space heating, water heating, and cooking require relatively low temperatures, and these activities dominate household energy consumption. In addition, many industries use low-temperature heat for processes that include drying seed crops, sterilizing food and medical equipment, boiling and distilling, and making paper and textiles. Most of the energy used to provide this heat is in the form of natural gas or electricity. Powering these activities with renewable energy poses a challenge, though much less of one than the high-heat activities discussed previously.

Low levels of heat can be supplied with electricity via electric heaters and heat pumps, by solar thermal collectors, and by geothermal sources. In addition, there is a very large potential role for the greater use of insulation and passive solar design in reducing the need for heating and cooling of buildings. Let's look at these strategies one by one.

Electric Heat

As noted in the previous section, the conversion of electricity to heat can be accomplished very efficiently; however, this use of electricity is usually not economically competitive. Nevertheless, in an all-renewable future where our dominant energy sources provide electricity directly, it may make sense to electrify space heating and cooking while reducing the need for energy use by designing buildings and systems for greater efficiency. One device that could help electrify more space heating economically is the heat pump.

A heat pump is a device—usually powered by electricity—that moves heat in the direction opposite of spontaneous heat flow by absorbing heat from a cold space and releasing it to a warmer one. The same device can often serve as an air conditioner and water heater. There are two main types of heat pumps used in buildings: ground-source and air-source. Ground-source heat pumps have been around longer and are more efficient, but they are typically more expensive to install. Air-source heat pumps have become widely available commercially only in the past few years, are relatively easy to install, and are typically much cheaper to operate than electric resistance heaters. Although some earlier models struggled to provide heat in below-freezing weather, many newer models are designed to operate to −18°C (0°F) or colder. Deploying heat pumps in superinsulated buildings (see the discussion of passive solar design later in this section) is one of the most easily identifiable, and most sensible, pathways for transitioning to an all-renewable energy system.

Solar Thermal

Solar water heaters are already common in many parts of the world. Indeed, according to the *Renewables 2014 Global Status Report* by the Renewable Energy Policy Network for the 21st Century (REN21), solar hot water capacity, at 330

GWh in 2013, was exceeded only by biopower generation in renewable energy capacity, with wind power a close third.[13] China leads with nearly 65 percent of solar thermal capacity followed by Europe with 20 percent.[14] In the United States and Canada, the principal application is for heating swimming pools. The main drawback with this energy source is simply that, especially in regions far from the equator, there are weeks or months during the year when solar heating has to be supplemented with gas or electric water heaters. Severe cold can create additional problems with freezing and breakage of pipes.

Solar plate collectors similar to ones used in water heaters can also be used to heat air. The use of water or air heated directly with sunlight (bypassing electricity) could be greatly expanded for some industrial applications, such as washing, sterilizing, drying, and baking, and in the paper and textile industries. The capital costs of deploying solar air and water heaters (and the energy embodied in them) would be far less than that entailed in photovoltaic panels or wind turbines used to make electricity, which then heats air or water. And in most applications, existing industrial machinery and distribution infrastructure could still be employed, with only the energy source needing replacement.

The potential for solar heat in industrial processes is perhaps even larger than that for households.[15] Where medium temperatures of 100°C to 400°C (212°F–752°F) are needed, solar concentrator technologies now used for electricity production could be adapted.[16] The main challenges would lie in either adapting production schedules to the sunniest times or finding ways to store thermal energy.

Geothermal

As discussed in chapter 3, high-quality geothermal energy is currently available only in regions where tectonic plates meet and volcanic and seismic activity are common, though low-temperature geothermal direct heat can be tapped nearly anywhere on Earth by digging a few meters down and installing a tube system connected to a heat pump (strictly speaking, this counts as stored solar energy).

Where geothermal energy is available and is close to an urban area, it can be used in district heating systems. Geothermal district heating (GeoDH) markets are developing throughout Europe, notably in Paris, Munich, and Hungary, with systems under consideration in the Netherlands, Madrid, and Newcastle (UK). It has been estimated that by 2020, most EU nations will have at least one GeoDH installation.[17] In the United States, the first GeoDH system was constructed in the 1890s in Boise, Idaho, but growth in this technology has been very slow. There are currently only twenty-one operating GeoDH systems in the country, with a combined capacity of about 100 MW thermal. China has deployed hybrid heat pump systems combining solar thermal with geothermal heat pumps.[18]

While the temperatures obtainable from geothermal sources are not high enough for industrial processes like steel and cement making, they are perfectly suited to low-temperature applications already discussed.

Insulation and Passive Solar Design

An important strategy in eliminating the need for fossil fuels to heat buildings will be to reduce the *need* for heating and cooling by designing and constructing buildings for greater energy efficiency. In Germany, thousands of structures have been designed according to the "passive house" (German: *passivhaus*) standard; these homes and commercial or public buildings typically require only a small fraction—often only 5 or 10 percent—of the energy used to heat and cool similarly sized conventional structures.[19]

Passive solar heating design entails three primary features: glazing for capturing sunlight, Trombe walls and other ways of storing heat, and insulation to maintain relatively constant temperatures. Important factors include orientation of the long side of the building toward the sun and the appropriate sizing of the mass required to retain and slowly release accumulated heat after the sun sets. Other passive uses of sunlight in buildings include daylighting and even solar cooling.

Depending on the study, passive solar homes cost less than, the same as, or up to 10 percent more than other custom homes; however, even in the latter case the extra cost will eventually pay for itself in energy savings. A passive solar building can provide only heat for its occupants, not electricity, which is

The EchoHaven community in Calgary, Canada, includes energy-efficient homes, including with passive house and net-zero design. (Credit: [cc-by-nc-sa] David Dodge, via flickr.)

a different technology; but if these techniques were used on all new buildings in climates that require heating, passive systems could go a long way toward replacing space heat from fossil fuels. Incorporating passive solar technologies into the design of a new home is generally cheaper than retrofitting them onto an existing home. Passive solar buildings, in contrast to buildings with artificial lighting, also provide a healthier, more productive work environment.

Limitations of passive solar heating include geographic location (clouds and colder climates make solar heating less effective), and the need to seal the house envelope to reduce air leaks, which increases the chance that pollutants will become trapped inside. The heat-collecting, equator-facing side of the house needs good solar exposure in the winter, which may require spacing houses farther apart and using more land than would otherwise be the case.

Fossil Fuels for Plastics, Chemicals, and Other Materials

Some readers may judge it to be outside the scope of this book to discuss non-energy uses of fossil fuels as feedstocks for plastics, fertilizers, pharmaceuticals, pesticides, road-paving asphalt, sealing tar, paints, inks, dyes, lubricants, solvents, paraffin, synthetic rubber for tires, and so forth. However, fossil fuels employed for these purposes are still subject to depletion, and in most instances still ultimately contribute to greenhouse gas emissions or local environmental pollution. Thus, if we wish to properly conduct an "all-renewable future" thought exercise, we should take these nonenergy uses of coal, oil, and gas into account.

Total nonenergy use of fossil fuels (including the energy used to process the feedstock in products), amounted to approximately 808.6 million metric tons of oil equivalent (Mtoe) in 2012. This accounts for 7.4 percent of total fossil fuel consumption of 10,927 million metric tons of oil equivalent.[20] The products produced from these fossil fuels are essential to agriculture, transportation, health care, and manufacturing; thus in many instances their importance to society may be far out of proportion to their energy footprint.

With regard to each of these materials, we must ask two questions: (1) How substitutable are fossil fuels in its production? and (2) Can this material be substituted with something else not derived from coal, oil, or natural gas? The list of fossil-fuel-based materials is quite long, and the authors of this report did not have the time or resources to make an exhaustive investigation. What follows is a highly selective exploration of just a few examples.

Plastics

Conventional plastics consist of a wide range of synthetic or semisynthetic organic chemicals that are malleable and moldable, including polyester, polyethylene, polyvinyl chloride (PVC), polyamide (nylon), polypropylene, polystyrene, polycarbonate, and polyurethane. They are typically derived from fossil fuels using processes involving heat (usually supplied by more fossil fuels) to yield specific molecules with desired qualities.

In recent years the chemical industry has devoted increasing effort to the production of bioplastics made from biomass sources, including vegetable fats and oils, cornstarch, pea starch, and microorganisms. Some bioplastics are designed to biodegrade, so that they can be composted (though they often break down rather slowly). These alternative plastics are suitable for making disposable items, such as packaging, cutlery, bowls, and straws; they are also often used for bags, trays, fruit and vegetable containers, egg cartons, meat packaging, and bottling for soft drinks and dairy products. Thermoplastic starch currently accounts for half the bioplastics market. Other types include cellulose bioplastics (cellulose esters, including cellulose acetate and nitrocellulose and their derivatives, such as celluloid); polylactic acid (PLA), a transparent plastic produced from corn or dextrose; poly-3-hydroxybutyrate (PHB), a polyester produced by certain bacteria processing glucose, cornstarch, or wastewater; and polyethylene derived from ethanol.

Compared to conventional plastics, bioplastics require less fossil fuel for their production and introduce fewer net greenhouse emissions if they biodegrade. They also result in less hazardous waste than conventional plastics, which persist in the environment for centuries.

However, fossil fuels are often still used as a source of materials and energy in the production of bioplastics. In our current industrial agriculture regime, petroleum and natural gas are used to power farm machinery, to irrigate crops, to produce fertilizers and pesticides, to transport crops to processing plants, and to process crops. The production processes for bioplastics also require heat and fuels for machinery, and these are usually supplied by fossil fuels. Further, producing bioplastics as well as biofuels in large quantities could accelerate deforestation and soil erosion and exacerbate water shortages.

Bioplastics production capacity stands at 1.7 million metric tons per year, still a small fraction of the total production of all plastics globally, which in 2013 reached 299 million metric tons.[21]

Fertilizer

Nitrogen (ammonia-based) fertilizer is produced using the Haber–Bosch process, usually employing natural gas as feedstock—though China, the world's largest fertilizer producer, primarily relies on coal as feedstock. The importance of this supplement to modern industrial agriculture can hardly be overstated. Over a hundred million tons of nitrogen fertilizer are applied annually around the globe; without it, Earth's soil might not be able to provide over seven billion humans the food they now consume. Indeed, almost half of the nitrogen found in a typical human's muscle and organ tissue originated in the Haber–Bosch process.[22]

In fertilizer production, while natural gas or coal is typically used both as a source of hydrogen to bind with atmospheric nitrogen and as energy for the process, hydrogen can be derived from other sources, including hydrolysis from water using electricity; electricity could also power the production process. Thus renewable energy–based fertilizer is chemically feasible—though it would be more costly to produce (unless and until natural gas and coal prices soar far higher than their current levels).

It is also possible to substantially reduce or even eliminate chemical fertilizer application using organic agriculture methods. Crop rotation can help with maintaining nitrogen levels, and simply planting a cover crop after the fall harvest significantly reduces nitrogen leaching while cutting down on soil erosion. Meanwhile, introducing nitrogen-fixing leguminous crops into the rotation cycle replaces nitrogen. Other substitution strategies could include broader and more effective use of animal and human manures (as civilizations did for millennia before the advent of synthetic fertilizers).

Cleverly designed polycultures that don't require synthetic fertilizer can sustainably outproduce synthetic-fertilizer-dependent monocultures on small and large farms, when counting total combined yields.[23] Further, mixing crops and reconnecting crop and livestock production consistently makes more efficient use of land, nutrients, and energy. However, these strategies usually require more labor and more farmer expertise. Thus a renewable energy future will likely entail more expensive natural fertilizers, more farmers as a proportion of the overall population, and more locale-based education for farmers in the use of organic production methods.

Paints

The main functional constituents of paints include binders, solvents, and pigments. Most modern paints are entirely made of, or with the use of, fossil fuels. For example, typical binders include synthetic or natural resins, such as alkyds, acrylics, vinyl-acrylics, vinyl acetate/ethylene, polyurethanes, polyesters, melamine resins, epoxy, or oils. For water-based paints, the solvent is water; however, for oil-based paints, solvents may include alcohols, ketones, petroleum distillate, esters, and glycol ethers. Pigments fall into two categories: natural pigments, including clays, calcium carbonate, mica, silicas, and talcs; and synthetics, including engineered molecules, calcined clays, blanc fixe, precipitated calcium carbonate, and synthetic pyrogenic silicas. Hiding pigments, which make paint opaque and protect it from ultraviolet light, include titanium dioxide, phthalo blue, and red iron oxide. Exterior paints contain fungicides, again made from, and using, fossil fuels.

Recent years have seen a dramatically increasing market share for eco-friendly, low-volatile organic chemical, and organic paints, which do away with fossil fuel–based solvents. However, these are typically still latex paints that use synthetic polymers, such as acrylic, vinyl acrylic, and styrene acrylic, as binders. Truly organic latex paints—in which the binders are produced from plant-based biochemicals, and fossil fuels are not used to power the production process—are feasible, though the authors were unable to confirm that any company currently produces them. For examples of paints that are truly free of fossil fuels, it is probably necessary to look to the era prior to organic chemistry. Paints made from organic flaxseed and linseed oil were used for centuries in Europe, while milk paint—made with milk protein (called casein) and lime—was the interior paint of choice in colonial America.

Asphalt

Nearly all paved roads are currently built using asphalt, a sticky, highly viscous or semisolid form of petroleum. A single kilometer of roadway typically requires roughly 320 barrels of oil (most of it in the form of asphalt) for its construction.[24] It is difficult to imagine how modern industrial society could operate without its ubiquitous road systems, yet our current roads are made from a depleting, nonrenewable material; contribute to climate change; and release toxic gases both during the construction phase and throughout their lifetimes.

Paving with concrete is an obvious alternative, and it is much longer lasting. However, we have already explored the energy intensity of current cement production and the difficulties likely to be encountered in redesigning the process to use renewable energy.

An alternative road-building material has been proposed in the form of a "sandstone" road surface produced by combining sand with a specific type of bacteria. Designers Thomas Kosbau and Andrew Wetzler won the green design competition in the 2010 Incheon International Design Awards (Incheon, South Korea) for their proposed biological substitute to asphalt, claiming that it could be produced at lower cost, while offering similar performance characteristics as a paving material. Sand is spread and compacted on a road surface then sprayed with a *Sporosarcina pasteurii* (formerly *Bacillus pasteurii*) solution; the microbes act to bind the sand into a tough material that is intended to sustain heavy traffic.[25]

Kosbau and Wexler's alternative road-paving material gained some publicity in 2010. The authors of this book were unable to discover whether or where their invention has undergone demonstration and performance testing on roads, but the same approach of using the microbial binder has since been tested for making building blocks and bricks[26] and is being investigated by NASA for use in constructing building blocks on Mars.[27] Until real-world commercialization occurs, it would be unwise to consider the "renewable paving" problem solved.

It is worth noting that, in a world of reduced mobility, the need for paving may also be significantly reduced.

Lubricants

Only a small proportion of overall petroleum production is diverted to the making of lubricants, but without these substances all the machines in the world with moving parts would soon grind to a standstill (friction is as unavoidable as corrosion). The properties of petroleum lubricants (high temperature stability, low viscosity breakdown, oxidative resistance) are impossible to match with vegetable oils.

Biobased lubricants are currently available on the market, but their man-ufacturing process is largely unclear owing to proprietary formulations. Most are based on vegetable oil–based stock, which can be chemically, thermally, or structurally altered to improve performance, and all contain a range of addi-tives, such as detergents, pour point depressants, viscosity index improvers, and rust inhibitors, to match the performance of petroleum-based lubricants in this area. Nonetheless, owing to the fundamental chemistry of vegetable oil–based stock, biobased lubricants remain deficient in terms of storage stability, thermal oxidative stability, low-temperature properties, corrosion protection, and hy-drolytic breakdown, thus reducing their potential scope of application.[28]

*   *   *

In this section we have surveyed only five examples of nonenergy uses of fossil fuels; however, they are crucial ones. They suggest that, in principle, current materials that rely on fossil fuels for feedstocks can be substituted, though often with a sacrifice in terms of higher cost or reduced functionality. More research and development, as well as wider commercial deployment, are needed. It also bears noting that, in instances where materials themselves have fossil fuel-free substitutes, the production processes for these substitute materials often em-ploy fossil fuels for transportation or as a heat source.

Summary: Where's Our Stuff?

This chapter discussed some easier and some harder ways to eliminate fossil fuels. The easier ways include heating buildings with air- and ground-source heat pumps, and heating water with solar collectors. The harder ways include making metals, cement, plastics, fertilizers, and roads without oil, coal, or natu-ral gas.

As we noted, electric kilns and solar furnaces already exist, though they are not currently used for large-scale production of pig iron or cement. However, using these technologies to produce what amounts to the material scaffolding of our industrial society would probably entail much higher costs than industry is accustomed to. Further, each industrial sector faces enormous costs in process redesign and in construction of new facilities to enable the use of renewable energy. These costs will be passed along in the prices of the output products.

We also touched upon the fact that current manufacturing processes for solar and wind energy technologies depend on high-temperature industrial processes currently fueled by oil, coal, and natural gas. Again, alternative ways of producing this heat are feasible—but the result would be higher-cost solar and wind power. We will further explore the current dependency of renewable technologies on continued fossil fuel consumption in the next chapter.

In principle, most of the problems we have identified are solvable—but at a cost and with serious questions regarding scale. A fully renewable energy future will entail higher costs for building and maintaining infrastructure, and the scale at which manufacturing can take place and infrastructure can be built using only renewable energy is highly uncertain. Without a massive mandatory program, the transition will take decades, and even with such a program we are likely to end up with a different kind of economy in which goods that incorporate metals—and infrastructure that involves steel and concrete—are more expensive and rare.

With plastics and chemicals, again substitution is possible in principle, at least in many instances. And once again the issues are cost, scale, and rate, along with tradeoffs (in some cases) of practical utility. There are no absolute barriers to a 100 percent renewable energy economy. But, as we have noted, it seems likely to be a smaller, slower, and more localized economy than our current one. Rather than a highly mobile consumer economy in which citizens are encouraged to buy as many goods as possible, and in which manufacturers pursue a strategy of planned obsolescence in order to encourage consumption, it will likely be one in which it is necessary to make goods that last longer, and to promote reuse and repair of older goods.

CHAPTER 6

Energy Supply: How Much Will We Have? How Much Will We Need?

T HERE IS SIMPLY NO WAY to accurately forecast exactly how much total energy is likely to be available in our 100 percent renewable future. There are too many variables at play—some technical, others economic or political. A few of the factors impacting future energy supply are favorable—including falling prices, technical improvements, and a generally favorable public attitude toward solar and wind. However, other factors that we have just surveyed pose challenges, including source intermittency, the need for storage and grid redesign, and the difficulties of electrifying heavy transport and many industrial processes. On balance, we believe the preponderance of factors support the assumption that energy quantities will be lower, perhaps significantly lower, than business-as-usual global energy demand projections from official agencies such as the International Energy Agency (which expects demand to rise 1.5 percent a year through 2035, nearly doubling over 2009 levels by 2050).[1]

This chapter explores in more detail why energy supplies are likely to be constrained in an all-renewable future, and then examines what this means. Some of the questions we'll address along the way include the following:

- Will we *need* less energy due to more efficient usage and the reduced need for energy conversion?
- Can we decouple energy growth from economic growth?
- Can we use renewable energy to build more renewable energy production capacity?

Perhaps the overall challenge of replacing all fossil-derived energy while continuing to grow the economic benefits of energy supplied to society can best be appreciated in historic terms. Humanity's past energy transitions (from wood to coal, coal to oil and natural gas) were driven by economic opportunity, not policy, and new types of energy were usually additions to, rather than replacements for, existing energy sources. Total supplies expanded quickly as the mix of energy sources evolved. However, what the world needs to do now is largely unprecedented—to force a rapid, policy-driven replacement of existing energy resources without sacrificing the benefits of the incumbent energy system, knowing that the characteristics of new energy resources may partially compromise the outcome.

The theoretical potential for solar and wind is vast. As noted earlier, the total amount of energy absorbed annually by Earth's atmosphere, oceans, and land masses from sunlight is approximately 3,850,000 exajoules—whereas humanity currently uses just over 500 exajoules of energy per year from all sources combined, an insignificant fraction of the previous figure.[2] If only 0.014 percent of the energy flow of sunlight could be captured, it would be enough to satisfy current world electricity demand. The potential is vast; however, as we have noted, limits are likely to be encountered in scaling up the technology required to harvest these enormous ambient energy flows, and in adapting current energy consumption patterns to using variable sources of electricity. And constructing enormous numbers of solar panels and wind turbines requires materials and energy, as well as financial capital.

The goal of this chapter is not to forecast future energy supplies quantitatively (as already noted, there are just too many variables), but rather to explore some additional factors that will impact levels of supply—and also to explore the relationship between energy supply levels and economic growth.

Energy Returned on Energy Invested of Renewables

To know how much useful energy we will have in an all-renewable world, we have to adjust assumed gross energy production figures by subtracting the energy invested in energy-producing activities. This tells us the *net energy* available to do useful work (as discussed in chap. 1). If the *energy returned on energy invested* (EROEI) ratio for the future renewable energy system (including not just panels and turbines but storage technologies and grid enhancements as well) is significantly lower than that of our current energy system, then even if total energy production stays the same, the amount of *useful* energy will decline.

When considering EROEI study results, it is helpful to keep two threshold numbers in mind. Charles Hall and others argue that returns above 3:1 are needed for an energy resource to be viable, while society needs a much higher overall EROEI (above 7:1) to support energy-consuming activities like education, health care, research, and the arts.[3]

Unfortunately, although EROEI studies are key to the economic evaluation of energy sources, the status of the net energy literature is far from satisfactory. Differences in methodology tend to yield widely ranging EROEI estimates for the same energy source. For example, EROEI studies of wind power have yielded results varying from 1.27:1 in a 1983 German study to 76.92:1 in a Danish study in 2000.[4] (In fairness, the technology and economics of wind power changed significantly between 1983 and 2000.)

EROEI is a system-level evaluation of a particular energy production pathway embedded in a specific industrial/economic network. Still, when the inherent complexities of the discussion and methodological differences are accounted for, it seems clear that some renewable energy production pathways have a much lower EROEI than those for most commercial fossil fuels. Indeed, the EROEI of some renewables is too low for them to serve as viable, self-sustaining energy sources; this is almost unquestionably the case for corn-based ethanol production in the United States, for example.

There is some controversy as to whether solar photovoltaic (PV) systems also have too low an EROEI to power industrial societies. A study by Marco Raugei concludes that the EROEI of PV technologies ranges from 19:1 to 38:1[5]; if these numbers are verified, then PV systems should easily be able to provide energy to operate industrial societies while in energy terms also "paying for" their own production and maintenance. However, a comprehensive operational study conducted by a pioneer of EROEI research (Charles Hall) and the manager of several of Spain's largest industrial PV power-generation facilities (Pedro Prieto) came to starkly different conclusions.[6] Prieto and Hall calculate an EROEI for Spanish PV of 2.4:1 to 7:1, depending on boundaries chosen. Graham Palmer arrives at similar results in his EROEI analysis of PV in Australia.[7] If verified, the Prieto–Hall and Palmer estimates would be very discouraging for the energy transition. However, it should be noted that the Prieto–Hall study has been criticized for its methodological inconsistency with other studies.[8] The authors start with a project-level analysis (of a single panel or PV farm, which would produce a full life cycle energy profitability metric) but then switch to an analysis of the entire PV industry in Spain for a given year without discussing the implications of the switch—that is, that energy-flow analysis is dynamic and factors like industry growth rate will impact the result. Critics of the Prieto–Hall study argue that it makes little sense to compare a flow-based EROEI with a full life cycle EROEI without correcting for the growth rate of the industry, which was not done.[9] A more recent meta-analysis by Bhandari et al. suggests a range of EROEI for PV of 8.7 to 34.2, depending on the technology and its siting.[10] These figures are generally supportive of Raugei's results and, while lower than the energy return numbers for conventional fossil fuels during their heyday, are still high enough to support an industrial economy.

The EROEI of wind has been the subject of less controversy, with a meta-analysis of fifty studies suggesting a likely value of 19:1 for systems in place.[11]

Examination of the EROEI of energy *sources* per se may not give us an accurate view of the energy costs associated with different energy *systems*. If electricity storage and redundant capacity are required to buffer the intermittency of

solar and wind, then these systemic energy costs need to be taken into account
as well. A study by Weissbach et al. showed that the EROEI of a solar or wind
energy system is reduced roughly by half when energy storage is added to the
analysis.[12] This confirms a conclusion many energy analysts have already ar-
rived at on the basis of economic calculations alone: that in designing renewable
energy systems it is preferable to minimize the need for storage and redundant
capacity through demand management wherever possible.

We are still at too early a stage in renewable energy deployment to know
how much storage and capacity redundancy will be needed, and we are at too
early a stage in EROEI studies to be able to judge whether the more optimistic
or more pessimistic results for PV are more accurate. However, if it turns out
that high levels of storage are required and that the middle-of-the-road EROEI
figures for solar PV of 10:1 and for wind of 19:1 (without storage) are justified,
then as society transitions away from high-EROEI fossil fuels its overall economic efficiency may decline, as a somewhat higher proportion of produced
energy will have to be reinvested into further energy production. This may have
implications for the possibility of further economic growth, as we will consider
later in this chapter.

Building Solar and Wind with Solar and Wind

The rapid build-out of renewables constitutes an enormous infrastructure project that will itself consume significant amounts of fossil-fuel energy (fig. 6.1).
While it is possible to imagine a solar panel or wind turbine factory operating
solely on electricity supplied by renewable electricity, it is much harder to envision entire supply chains—from the mining of ores to the final delivery and
installation of panels and turbines—functioning without fossil energy, at least
in the early stages of the transition.

As we saw in chapter 4, fossil fuels are currently used for mining raw materials, constructing roads and factory buildings, and transporting raw materials
and finished products. Theoretically solar and wind technologies could supply
the energy for these processes, using electric mining, manufacturing, and hauling equipment (perhaps, for example, electricity could be produced on-site and

transmitted via cables to mining equipment). Fossil fuels are also used to supply high levels of heat for extruding aluminum, making copper wire and plate, and producing iron and cement.[13] Solar and wind electricity can in principle produce high heat for these purposes. However, as discussed in chapter 5, it would be much more expensive to generate the temperatures needed with electricity from solar panels or wind turbines than from burning fossil fuels. This would add to the cost of renewable energy technologies. To the authors' knowledge, no real-world pilot projects exist in which all the industrial processes involved in making renewable energy technologies are powered by renewable energy.[14]

A bootstrap transition scenario (in which renewables provide the energy needed to build more renewables, while still supplying much of the rest of the energy that society needs) seems daunting in principle. Where *will* the energy for the transition come from, then? Realistically, most of it will have to come from fossil fuels—at least in the early-to-middle stages of the process. And we

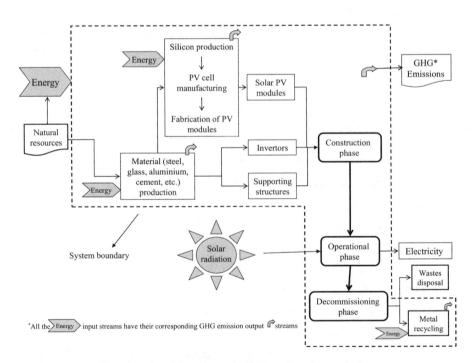

Figure 6.1. Considerations in a life cycle analysis of a solar photovoltaic system.
Source: R. Kannan et al., "Life Cycle Assessment Study of Solar PV Systems: An Example of a 2.7 kW Distributed Solar PV System in Singapore," *Solar Energy* 80 (2006): 555–63.

will be using fossil fuels whose economic efficiency is declining due to the ongoing depletion of existing stocks of high-quality oil, gas, and coal. Again, this implies higher overall costs. But using only renewable energy to build renewables would be slower and even more expensive.

The faster we push the energy transition, the more energy will have to be diverted to that gargantuan project, and the less will be available to all the activities we're already engaged in (running the food, transportation, manufacturing, communications, and health care sectors, among others). Moreover, a faster transition will delay the point at which large amounts of useful net energy are available from newly installed renewable energy generators.

If fossil fuels will be required for constructing solar panels, wind turbines, and the infrastructure that enables us to use them, then high build-out rates for renewable energy technologies may have implications for carbon emissions.[15] The faster we push the transition, the higher the emissions—unless we rapidly curtail *current* uses of fossil fuels in the meantime (reducing fossil energy consumption faster than it can be replaced by renewable energy), implying a reduction in energy consumption and therefore in gross domestic product (GDP).

Investment Requirements

A realistic assessment of future energy availability would also have to take into account the requirement for financial investment capital. While solar and wind have enjoyed rapidly increasing rates of installation during most of the past decade, transition plans envision an even more rapid shift, involving much higher levels of investment in generation capacity, storage, grid upgrades, and transport alternatives. Will sufficient money be available?

The affordability problem is finessed in some published energy transition studies. For example, in a recent plan for a conversion of the U.S. economy to 100 percent renewable energy by 2050, Mark Jacobson et al. count savings from avoided costs of climate change and health damage in their estimate of the affordability of such a comprehensive and rapid conversion.[16] However, as discussed earlier, avoiding externalized costs associated with fossil fuel consumption might make the renewable energy transition more affordable on a society-wide basis, but that does not actually mean the transition will be affordable on its own terms.

Estimating how much a total energy transition would cost is difficult. The problem is simplified greatly by including only the direct cost of solar panels and wind turbines, but doing so is unrealistic. Actual costs would include required investments in new technology for the transportation, agriculture, and manufacturing sectors; in new equipment for building operations, and for energy efficiency retrofits; in grid redesign; in energy storage; and in redundant generation capacity.[17] For the average American household, costs for installing insulation, an air-source heat pump, an electric stove (assuming they currently have a gas stove), and a solar water heater with on-demand electric water heater backup would run into many thousands of dollars; this does not include the cost of an electric car (we assume the average family will be trading out its current car at some point anyway) or solar panels and batteries (our hypothetical family may choose to buy grid-supplied renewable electricity). Just multiplying these outlays by the number of American households yields figures in the hundreds of billions of dollars, but this does not include the far greater costs to utilities, or the research and development and retooling costs in the energy-consuming industries just mentioned. In a *Scientific American* article in 2009, Mark Jacobson estimated the total cost of the transition at about $100 trillion, spread over 20 years.[18] However, this includes primarily energy supply requirements and excludes the necessary investment in revamping all economic sectors on the consumption side. The latter could easily match the necessary investment in energy supply.

Actual rates of investment in renewable energy globally have leveled off in the past four years (fig. 6.2), with investment rates in North America and Europe stalling or shrinking while China continues to surge ahead.

In 2014 the world's nations invested $270 billion in renewable energy (92 percent of that was for solar and wind), which represented roughly one-sixth of all energy spending.[19] Overall, investments in conventional fossil fuel production continue to dominate.

Jacobson's estimate of the energy supply cost of the transition ($100 trillion over 20 years) amounts to $5 trillion per year in required investment. With

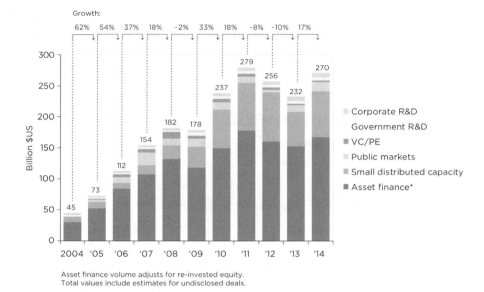

Figure 6.2. Global new investment in renewable energy by asset class, 2004–2014.
Source: Frankfurt School–UNEP Collaborating Centre for Climate & Sustainable Energy Finance (FS-UNEP) and Bloomberg New Energy Finance, Global Trends in Renewable Energy Investment 2015–Chart Pack (Frankfurt: FS-UNEP, 2015), http://fs-unep-centre.org /sites/default/files/attachments/unep_fs_globaltrends2015_chartpack.pdf.

the current investment rate stuck at around $270 billion per year, it is clear that rates of investment will have to increase by a factor of more than 10 if we are to come close to supplying sufficient energy from renewables to replace all current energy supplied by fossil fuels. The world currently spends $1.8 trillion annually on military activities, so the required investment rate should not be ruled out as unrealistic in principle.[20] However, the scale of what is needed is breathtaking.

Funding for enormous new infrastructure spending projects is difficult to organize unless economic and financial systems are stable and expanding. One of the authors of this book has argued elsewhere that three converging factors (too much debt, rising energy costs, and increasing environmental stress) are leading to the end of economic growth as it was known during the latter half of the twentieth century.[21] Real economic growth has indeed slowed in the world's wealthy industrial nations in the past couple of decades.[22] Since the 2008 crisis, central banks have deployed low interest rates and quantitative easing, and governments have bailed out banks and major industrial firms while engaging in deficit-funded stimulus spending. In theory these actions should have produced a robust recovery, but the result has instead been more commodities, stock market, and real estate bubbles—with almost all the benefits going to society's wealthiest.[23] Further, the marginal productivity of debt—the amount of additional GDP produced by one dollar more of debt—has plummeted from around $3.00 in the 1950s to near zero today, indicating that debt is no longer providing the economic boost it did in the past.[24] Meanwhile, real living standards in the United States and much of Europe drift lower.[25] A fairly robust literature is developing to attempt to account for this "secular stagnation," which many economists now think could continue for decades.

When economic growth ceases, as it does in times of recession, investment capital tends to become scarce. Thus scarce investment capital could pose a barrier to a robust renewable energy transition. The Keynesian solution for recession is for governments to become the borrowers and spenders of last resort in order to prime the growth pump. Could governments and central banks, following the Keynesian formula, simply print the money needed to fund the energy transition? This is just one of many currently unanswerable questions we are likely to encounter along the path toward a renewable future.

The Efficiency Opportunity: We May Not Need as Much Energy

In the production of electricity from coal and natural gas, about 60 percent of the primary energy contained in the fuel is lost in the conversion process.[26] Solar and wind electricity sources do not require a conversion process and therefore do not incur these high losses. This amounts to a substantial amount of potential energy savings: out of 197 billion gigajoules of primary energy currently flowing to the entire global electricity sector, 117 billion gigajoules wind up as conversion losses; this is energy that will no longer be needed in an all-renewable future.[27]

In addition, electric motors are significantly more efficient than internal combustion engines. While the latter are only 20 to 30 percent efficient (with most of the energy contained in gasoline lost as waste heat),[28] electric motors can be 92 percent efficient at translating energy into motive force.[29] Thus the more we electrify transportation and other uses of combustion engines, the less energy we will need in order to produce the same economic and social benefits. This has practical implications for the energy transition. In the United States, passenger vehicles currently use about as much energy in the form of gasoline as is consumed in the entire electricity sector. But transitioning to electric cars would not require a doubling of electricity generation; we could do it with about 29 percent additional electricity.[30]

As discussed in chapter 5, a great deal of energy could also be saved in space conditioning if all homes and buildings had passive-house levels of efficiency. Assuming a generous 90 percent cut in energy use for this purpose, consider another 350 million metric tons of oil equivalent (Mtoe) or so energy saved.[31]

Of course, to obtain a realistic estimate of overall energy savings we should also consider some *inefficiencies* that an all-renewable energy system might bring with it. One of these is tied to storage: storing a ton of coal or a gallon of gasoline implies little direct loss (though there are costs for the tanks and other storage infrastructure), while electricity storage always involves losses. The percentage of electricity that would be lost in storage annually in an all-renewable future would depend on a range of factors, including the types of storage used and the degree to which storage is used to buffer intermittency (as opposed to using capacity redundancy or demand management for this purpose). Also, if grids were expanded to enable load balancing over longer distances, this would entail higher electricity transmission losses. Still, on balance, there are very large opportunities for energy savings, though many of these would take time and substantial investment to realize. An electrified, optimally efficient society might need only half to two-thirds of current primary energy consumption to yield similar economic benefits.

All published renewable energy transition scenarios highlight this opportunity for obtaining equal economic benefits from reduced primary energy consumption. Most go further and assume that even greater reductions in energy use can be achieved while still supporting economic growth. But this assumption is controversial, as we are about to see.

Energy Intensity

Historically, there has been a close correlation between energy use and economic activity (see fig. 2.2). Increased energy consumption is associated with economic growth; during times of economic recession, energy consumption often declines.[32] This correlation makes sense, as everything we do requires expenditure of energy. Policy makers do not want to sacrifice prospects for economic growth in order to curtail fossil energy sources in favor of solar and wind. Yet there are good reasons to conclude that the energy transition will leave us with less useful energy than historic trends would lead us to expect. Is it possible to stretch the link between energy consumption and GDP growth so as to have more of the latter with less of the former?

Energy intensity (measured as the ratio of the consumption of final energy, meaning usable forms of energy such as heat or electricity, to GDP) varies from nation to nation.[33] There is evidence that the energy intensities of both the United States economy and the global economy have indeed been falling (fig. 6.3),[34] though a recent study by Wiedmann et al. suggests that historic "decoupling" of economic growth from increased energy usage has been significantly overstated.[35] The reasons for energy intensity improvements are summarized as follows in a paper by Jesse Jenkins and Armond Cohen:[36]

1. *Sectoral shifts* in the composition of the global economy, such as the increasing importance of services as a share of global GDP, which tend to expend much less energy per unit of economic activity than heavy industry or agriculture;
2. *Substitution* of other economic inputs for energy, such as an increased reliance on capital or labor in productive processes in lieu of energy inputs;
3. Improvements in *primary to final energy conversion efficiency*, or the efficiency at which primary energy supplies, such as coal, oil, or renewable energy inputs, are converted to usable, final forms of energy such as heat or electricity;
4. Improvements in *end-use energy efficiency*, or the amount of final energy inputs needed to deliver a given energy service, such as heating, cooling, transportation, or industrial process energy inputs.

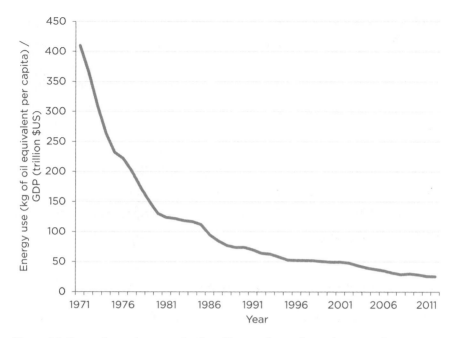

Figure 6.3. Energy intensity per unit of world gross domestic product over time.
Source: World Bank, World Development Indicators, http://data.worldbank.org/data
-catalog/world-development-indicators.

Is there reason to think energy intensity can be reduced significantly as we
transition to renewable sources? The energy savings from slashing energy conversion losses (no. 3 in the preceding list) and from the replacement of combustion engines with electric motors (no. 4 in the preceding list) discussed in the
previous section would almost certainly drive considerable further improvement in energy intensity. But these strategies have limits.

In a review of seventeen published decarbonization scenarios, Loftus et al.[37]
found that all of the scenarios rely upon improvements in energy intensity that
are unprecedented in history.[38] Three scenarios that exclude nuclear and carbon
capture and storage technologies (i.e., the ones that depend almost entirely on
growth in wind and solar power) require the fastest energy intensity improvements. The authors also noted that all of the studies they surveyed offer little
detail on how to decarbonize the industrial and transportation sectors, and on
needed energy system transformations." Loftus et al. conclude with the following comment, with considerable relevance for this book: "To be reliable guides

for policymaking, scenarios such as these need to be supplemented by more detailed analyses realistically addressing the key constraints on energy system transformation."

A 2014 report by PriceWaterhouse Coopers notes that the decoupling of emissions growth from economic growth has averaged only 0.9 percent per year since 2000.[39] This raises questions about the prospects for meaningful reductions in energy intensity beyond what can be achieved by reducing conversion losses and replacement of combustion engines with electric motors. Industry already has a cost-cutting incentive to improve efficiency; policy makers may have limited ability to increase the rate of efficiency improvements above this "exogenous" background rate.

The Role of Curtailment and the Problem of Economic Growth

If we won't have as much energy, and we can't improve efficiency at a continuous and dramatic rate—and therefore energy intensity cannot be reduced at unprecedented rates—then the economy will likely shrink. Rather than merely streamlining economic activities, we will have to curtail them, at least to a certain degree. Perhaps aviation offers the most pertinent example: as we have seen (in chap. 4), there are no easy or inexpensive substitutes for kerosene-based jet fuels, and so it is difficult to imagine the continued growth of this industry as carbon-based fuels are fairly quickly eliminated. Altogether, it is difficult to avoid the conclusion that an all-renewable future will offer less economic growth, no growth, or negative growth. But then again, the world is already seeing a reduction in economic growth rates. Since fossil fuels are finite, they cannot fuel perpetual growth in any conceivable instance. Thus it would be specious to argue that we face a choice between renewable energy and reducing greenhouse gas emissions on one hand, and economic growth from continued reliance on fossil fuels on the other.

A few climate scientists have already suggested that dealing with global warming could have serious implications for the economy. Kevin Anderson and Alice Bows of University of East Anglia's Tyndall Centre for Climate Change Research have calculated that a "carbon budget" consistent with a threshold of

2°C would entail an 8 to 10 percent annual reduction of emissions in industrialized nations, which would be, in Anderson's words, "incompatible with economic growth."[40] The 1.5°C goal referenced in the international agreement reached in Paris in December 2015[41] makes the challenge of achieving growth while massively reducing greenhouse gas emissions even more daunting.

The Intergovernmental Panel on Climate Change's (IPCC's) Fifth Assessment Report (2014) admits the difficulty of the renewable energy transition in this regard. "No single mitigation option in the energy supply sector will be sufficient," the report warns.[42] To stabilize the climate at an average global surface temperature no higher than 2°C above the preindustrial level, scenarios relying almost entirely on solar and wind energy would, in the opinion of the IPCC report's authors, require global energy supply to be radically curtailed below currently projected demand.[43]

Again: we cannot estimate how much energy will be available in an all-renewable future, other than to suggest that it will probably be significantly less than business-as-usual demand projections. Thus the energy transition constitutes an important challenge not just for scientists and engineers but for economists and policy makers as well. How shall we maintain social and material benefits to the world's people as population continues to grow, but energy availability declines and economies stall and contract?

The tapering of economic growth really should come as no surprise: a long-standing school of thought says that physical expansion cannot continue forever on a finite planet.[44] However, tapering presents serious challenges not just for political and economic systems but for the renewable transition itself: how are societies to obtain sufficient funding for the rapid and dramatic expansion of renewable energy infrastructure if their economies are stagnant rather than growing? Perhaps the worst outcome of all would come from a failure to plan for economic tapering: in that case, societies would deploy futile strategies to restart growth, while frittering away opportunities to prepare for a renewable, postgrowth future.

Of course, the fossil fuel lobby uses fears of economic hardship as an excuse to say that the energy transition should be delayed as long as possible. However, the reverse is true: the longer the transition is delayed, the more expensive and perilous it will become. The world's remaining high-quality and inexpensive-to-produce fossil fuels are depleting rapidly, so if a transition to alternative energy sources is *not* organized rapidly, economic contraction will still result. But in that case, we eventually end up with catastrophic climate change and *no* viable energy system.

As many economists have pointed out, GDP growth is a poor indicator of societal progress or well-being. For example, if power plant emissions are reduced due to the expansion of renewable energy, this could result in a decline in hospital stays and drug prescriptions related to asthma attacks, and this would in turn lead to lower GDP, even though it reflects an improvement in well-being. Extreme storms damage buildings, which then need to be repaired, increasing GDP—but well-being has of course declined in the process. GDP is the sum of all consumption in the economy (household, business, and government, along with net exports); thus we measure our well-being by how much we consume, and we have trapped ourselves into believing that this quantity must increase year after year.

Replacing GDP with a more robust and realistic measure of economic success is just one of the tactics proposed by postgrowth economists, such as Peter Victor, who recognize that the rapid expansion of population and consumption that characterized the twentieth century will inevitably subside in the decades ahead.[45] Victor and others propose ways to promote full employment and higher quality of life as consumption of energy and materials declines.

CHAPTER 7

What About ... ?

THIS BOOK IS ESSENTIALLY a thought exercise designed to explore some of the issues involved in transitioning our economy to 100 percent renewable energy. Some readers may chafe at the boundaries of this exercise. Why rely so much on wind and solar, rather than envisioning a more diverse mix of low-carbon energy sources? We chose our framework because it was simple and clear, and because this is a future that is indeed being widely proposed. The state of Vermont, for example, has announced the official goal of sourcing 90 percent of all its energy (not just electricity) from renewable sources—mostly solar and wind—by 2050. Moreover, studies have been published purporting to show that a 100 percent wind, solar, and hydro energy regime is both possible and affordable,[1] and prominent climate-oriented environmental organizations are now calling for that goal.[2] Further, we ourselves believe that a full transition to renewables is necessary and achievable, provided society is willing to accept adjustments, both profound and minor, to the ways it uses energy.

As we have seen, relying entirely on renewable energy entails some hefty challenges. We have discussed at some length the problem of source intermittency and the need for energy storage, grid redesign, and capacity redundancy;

the environmental and land use challenges of installing very large numbers of solar panels and wind turbines; electrification and the revamping of energy-consuming equipment; and the requirements for very high levels of investment. The conclusion we have reached so far is that, realistically, a mostly wind-and-solar future will likely provide less energy overall, less mobility, and less manufacturing capacity. This conclusion is likely to be unwelcome to many readers, again leading to objections regarding the study's narrow boundary assumptions. This chapter addresses three of the most likely of those objections.

Nuclear Power — non renewable

We cursorily explained our reasons for not including nuclear power in our "renewable future" energy mix in the introduction. The main reason is simply that nuclear fuel is not renewable. Some readers will nevertheless disagree with this decision, since (excluding mining, transport, enrichment, plant construction, and plant decommissioning) the nuclear fuel cycle generates no carbon emissions. For this reason there are many environmentalists and climate activists—including former National Aeronautics and Space Administration climate scientist James Hansen—who argue that nuclear has to play an expanded role as part of the energy transition away from fossil fuels. Therefore it may be helpful for us here to provide a more detailed discussion of nuclear power.

Nuclear electricity is reliable and relatively cheap (2.9 cents per kilowatt-hour) once the reactor is in place and operating. In the United States, while no new nuclear power plants have been built in many years, the amount of nuclear electricity provided has grown during the past 15 years due to the increased efficiency and reliability of existing reactors.

However, uranium, the fuel for the nuclear cycle, is a depleting resource. The peak of global uranium production rates is likely to occur between 2040 and 2050, which means that nuclear fuel is likely to become more scarce and expensive over the next few decades.[3] Already, the average grade of uranium has declined substantially in recent years as the best reserves have been depleted.[4] Recycling of fuel and the employment of alternative nuclear fuels are possible, but the technology has not been adequately deployed.

Nuclear power plants are so costly to build that unsubsidized nuclear plants are not economically competitive with similar-sized fossil-fuel plants. Government subsidies in the United States include those from the military nuclear industry, as well as nonmilitary government subsidies including artificially low insurance costs. New power plants also typically require many years for design, financing, permitting, and construction.

The nuclear fuel cycle entails substantial environmental impacts, which may be greater during the mining and processing stages than during plant operation, even when radiation-releasing accidents are taken into account. Mining entails ecosystem removal, dust, large amounts of tailings (equivalent to 100 to 1000 times the amount of uranium extracted), and radioactive particles leaching into groundwater. During plant operation, accidents causing small to large releases of radiation can impact the local environment or much larger geographic areas, potentially making land uninhabitable (as with Chernobyl and Fukushima).

Storage of radioactive waste is highly problematic. High-level waste (like spent fuel) is much more radioactive and difficult to deal with than low-level waste and must be stored on-site for several years before transferal to a geological repository. The best-known way to deal with waste, which can contain lethal doses of radiation for thousands of years, is to store it in a geological repository, deep underground. Yucca Mountain in Nevada, the only site that has been investigated as a repository in the United States, was ultimately rejected. More candidate repository sites will need to be identified soon if the use of nuclear power is to be expanded in the United States. Even in the case of ideal sites, over tens of thousands of years waste could possibly leak into the water table. The issue is controversial even after extremely expensive and extensive analyses by the Department of Energy.

Nearly all commercial reactors use water as a coolant. Heat pollution from coolant water discharged into lakes, rivers, or oceans can disrupt aquatic habitats. In recent years, a few reactors have had to be shut down due to water shortages, highlighting a future vulnerability of this technology in a world where droughts are becoming more common due to climate change. During the 2003 heat wave in France, several nuclear plants were shut because the river water was too hot.

Reactors must not be sited in earthquake-prone regions due to the potential for radiation release in the event of a serious quake. Nuclear reactors are often cited as potential terrorist targets and as potential sources of radioactive materials for the production of terrorist "dirty bombs."

Hall et al. reviewed net energy studies of nuclear power that have been published to date and found the information to be "idiosyncratic, prejudiced, and poorly documented."[5] The largest issue is determining what the appropriate boundaries of analysis should be. The review concluded that the most reliable energy returned on energy invested (EROEI) information is quite old (it showed an EROEI in the range of 5–8:1), while newer information is either highly optimistic (10:1 or more) or pessimistic (low or even less than 1).

The nuclear power industry is shrinking in most of the older industrial nations; only in China and India are substantial numbers of new reactors being planned. Hopes for a large-scale deployment of new nuclear plants rest on the development of new technologies: pebble-bed and modular reactors, fuel recycling in fast reactors, and the use of thorium as a fuel. However, each of these new technologies is problematic for one reason or another. The technology to extract useful energy from thorium is highly promising but will require many years and expensive research and development to commercialize. The only breeder reactors in existence are closed, soon to be closed, abandoned, or awaiting reopening after serious accidents: BN-600 (Russia, end of life 2010); Clinch River Breeder Reactor (United States, construction abandoned in 1982 because the United States halted its spent-fuel reprocessing program, making breeders pointless); Monju (Japan, potentially coming online again after a serious sodium leak and fire in 1995); and Superphénix (France, closed 1998). The ultimate technological breakthrough for nuclear power would be the development of a commercial fusion reactor. However, commercial deployment of fusion still appears to be decades away and will require much costly research.[6]

China now has about 20 GW of nuclear capacity online, with a target of about 70 GW by 2020 (a target it is likely to miss), and envisions about 250 GW by 2050. The nation is already a net importer of uranium. Chinese nuclear plans don't foresee alternatives to the standard uranium cycle, such as improvements to the nation's own native pebble-bed reactor design, until after 2035 at the earliest. Therefore, realistically, most nuclear power plants constructed in the short and medium term worldwide will be only incrementally different from current designs.

In order for the nuclear industry to grow sufficiently so as to replace a significant portion of energy now derived from fossil fuels, scores if not hundreds of new plants would be required, and soon. The enormous investment requirements for such a build-out would probably preclude a simultaneous large-scale build-out of solar and wind generators. But more realistically, given the expense and long lead time entailed in plant construction, the industry may do well merely to build enough new plants to replace old ones that are nearing retirement and decommissioning.

In short, we do not see a nuclear renaissance as a realistic alternative to a massive shift toward renewable energy in addressing the climate dilemma.

Carbon Capture and Storage

If stopping climate change is our main goal, isn't it possible to do this without completely phasing out fossil fuels by capturing and burying carbon emissions? That way, we could continue burning coal to generate cheap electricity (and use that electricity to power automobiles and an increasing share of industrial processes), while simultaneously reducing the release of carbon dioxide into the atmosphere.

For years, Americans have seen billboards and TV commercials touting "clean coal," while politicians from both major parties have extolled its promise. The technology to capture carbon emissions from coal-fired power plants has been tried and tested. Yet today almost none of the nation's coal-fueled electricity-generating plants are "clean."

Why the delay? The biggest problem for clean coal is that the economics don't work. Carbon capture and storage (CCS) is extremely expensive. That gives the power industry little incentive to implement it in the absence of a substantial carbon tax.

Why would implementing CCS be so expensive? For starters, capturing the carbon from coal combustion is estimated to consume 25 to 45 percent of the power produced, depending on the approach taken.[7] Add to this the energy costs for transport, injection, and storage management. The result would inevitably be not only higher prices for coal-generated electricity but also the need for more power plants to serve the same customer base. Other technologies designed to make carbon capture more efficient aren't commercial at this point, and their full costs are unknown.

Capturing and burying just 38 percent of the carbon released from current U.S. coal combustion would entail the manufacturing and installation of pipelines, compressors, and pumps on a scale equivalent to the size of the nation's oil industry, requiring tremendous energy expenditures.[8] And, although bolting CCS technology onto existing power plants may conceivably be possible, it would be exceedingly inefficient. A new generation of plants capturing carbon dioxide prior to coal combustion would do the job much better—but that means replacing roughly 600 current-generation power plants. Altogether, the U.S. Department of Energy estimates that wholesale electricity prices with the initial generation of CCS technology would be 50 to 80 percent higher than current coal-based power.[9]

Thinking long term, the economics of coal—and natural gas, for that matter—are likely to get worse, with or without the vast investment required for CCS implementation. After all, coal and natural gas are nonrenewable, finite in quantity, and therefore subject to depletion. Rates of production of coal from most regions of the United States are in decline. And as depletion forces the mining of lower-quality resources, production costs will rise because of the need for more-sophisticated extraction technologies. Declining output is inevitable sooner or later.

The only thing that keeps coal-based electricity cheap today in relation to power from renewable energy sources is the industry's ability to shift the hidden costs—environmental and health damage—onto society. If, as climate regulations inevitably kick in, the coal power industry adopts CCS as a survival strategy, the task of hiding from the market the real and mounting costs of coal can only grow more daunting. Any lingering economic advantage over wind and even solar will disappear.

On top of all this, CCS doesn't address the full range of coal's impact on society. It won't banish high rates of lung disease, because it doesn't eliminate all the pollutants from the combustion process or deal with the coal dust from mining and transport. It also doesn't address the environmental devastation of "mountaintop removal" mining.

This is not to say that clean coal has no future whatever. Coal plants with CCS will likely be built in situations where captured carbon dioxide can be used to generate extra income—for example, by using it to stimulate old oil wells or make cement. But even a dramatic increase in such uses would put only a small fraction of carbon from coal to work.

A full transition of today's coal power industry to CCS is extremely unlikely unless the economics substantially change for some currently unforeseeable reason. And other technological advances, like more-efficient coal-fired plants, can only slow the growth of harmful emissions at best.[10]

In all likelihood, the real future of carbon sequestration lies elsewhere—with reforestation and agricultural methods that build topsoil. Atmospheric carbon levels are currently at 400 parts per million (ppm) of carbon dioxide, while a consensus has emerged that a "safe" level would be below 350 ppm. One ppm is equal to 2.125 gigatons (Gt) of carbon; thus we need to safely sequester 106.25 Gt of carbon in order to return to a safe climate regime. Is there sufficient potential absorptive capacity in forests and soils to accomplish this?

Society has removed 136 Gt of carbon from soils through agriculture and land use. There is the potential to reverse the trend by minimizing tillage, planting cover crops, encouraging biodiversity, employing crop rotation, expanding management-intensive pasturing, and introducing biochar to soils.[11]

Deforestation has also contributed significantly to the historic increase in atmospheric carbon dioxide. It makes sense therefore that reforestation could diminish atmospheric carbon. Unfortunately, climate change is putting pressure on forests, even as we want them to recover. Nevertheless, a recent study shows large regional potential for sequestration, especially in the tropics.[12]

Massive Technology Improvements

Some readers may feel that we have failed to take into account the possibility of extraordinary new developments in energy research. In the computing world, Moore's law describes a trend of rapidly declining cost and increasing functionality regarding transistors. During recent decades, the number of transistors that can be crammed into a square inch of integrated circuit has doubled approximately every two years. Memory capacity, computer processing speed, and the number of pixels in digital cameras have shown the same trend. Why shouldn't renewable energy technology achieve a similar pace of improvement in output and efficiency?

Surely we can and should expect improvements to solar panels and wind turbines, and technological refinements are in fact occurring. As just one example, translucent photovoltaic modules are now feasible.[13] However, there are inherent physical limits to all processes and materials. Microprocessors have offered a unique opportunity for rapid technological advancement that may not be replicable in other fields. Areas of technology that involve massive infrastructure that is expensive to build and replace understandably evolve more slowly. Our energy infrastructure is in that category.[14]

What about entirely new energy resources? News reports occasionally inform us of experiments with cold fusion that purport to show high levels of anomalous energy output; or with artificial photosynthesis, which promises to be far more efficient than natural photosynthesis.[15] Couldn't the development of one or both of these technologies constitute a "black swan" event capable of changing the energy game overnight?

It's unlikely. In any case, it would have been pointless for us to try to factor black swans into our future energy scenarios. We don't know what the actual costs for these possible future energy devices would be, nor do we know their scalability or their EROEI, so no useful analysis is possible.

Even in the best case, it will take time to get from the point of discovery of a new energy process to the commencement of build-out of commercial devices. During this period, perhaps a decade or two, development and testing of products would occur. The build-out of those products to replace current energy production technology would likely take even longer, probably another two or three decades.

Vaclav Smil, author of *Energy Transitions: History, Requirements, Prospects*, tells us that an energy revolution takes 40 years at minimum.[16] Since we will need to have the renewable energy revolution largely completed 40 years from now in order to avert catastrophic climate change, that means we will have to count mostly on technologies that have already passed through the research and development stages. Solar and wind have done so; supporting policies (such as feed-in tariffs) have been tested; and investment capital is already flowing toward the build-out of these technologies. Substantially different and more efficient energy technologies may emerge later this century, but for the foreseeable future the fates of our economy and of the global climate appear to hang largely upon the success or failure of our adoption of solar and wind power, and on a wide range of technological adaptations to intermittent energy at lower overall levels of supply.

Much the same must be said for massive efficiency improvements in energy consumption technologies. We have tried to identify and factor in the advantages of existing technologies such as air-source heat pumps, LED lighting, passive-house design, electric cars and bicycles, and public transit options like streetcars and light rail. We did not attempt to estimate the likely contribution of technologies at a very early phase of adoption, such as 3-D printing and the "Internet of Things" (though the latter is discussed briefly in chap. 3).[17] Though extravagant claims have been made for how these technologies could reduce the need for product transportation and increase energy efficiency, there simply isn't enough real-world data to tell whether such claims are realistic or overblown.

*   *   *

In a nutshell, there is good news and bad news for society's efforts to transition away from reliance on fossil fuels and to instead adopt renewable energy technologies.

The good news is that most of what we currently do with fossil fuels can be done with renewables: solar and wind can generate electricity, cars can be battery powered, solar thermal can heat water for our homes (at least during sunny periods), biofuels can power heavy transport. It is possible to use solar concentrators or hydrogen (produced from renewable electricity) to run high-heat industrial processes, and biofuels could conceivably even power ships and a much-reduced global fleet of airplanes.

The bad news is that some of these substitutions will be very expensive, some will not scale up easily, and most will require considerable research and development. In some cases, higher investment requirements will probably be ongoing as a result of higher materials and process costs.

Few options constitute direct drop-in replacements for current technologies. This means we will not only have to improve energy production technologies and scale them up rapidly, but we will also need to replace technologies that use energy if no drop-in substitute is available. This will require time and money. For example, replacing every gas stove in the United States with an electric stove will entail a nontrivial expense. And gasoline cars have an average lifetime of twenty years, so electrifying the automobile fleet will take time. No plane that runs on hydrogen currently exists. For every supply-side cost, we are likely to see a consumption-side cost as well.

This may be a good place to reemphasize the fact that only about 20 percent of the energy we use daily is in the form of electricity. That means 80 percent of all energy services today need to be electrified or we need to find a renewable alternative, preferably a drop-in substitute, requiring massive research and development expenditures for developing both the substitutes themselves as well as new process technologies. This may also a good place to point out once again that half of the energy we use today is essentially "wasted."

We citizens of industrialized nations will have to change our consumption patterns. We will have to use less overall and adapt our use of energy to times and processes that take advantage of intermittent abundance. Mobility will suffer, so we will have to localize aspects of production and consumption. And we may ultimately forgo some things altogether. If some new processes (e.g., solar or hydrogen-sourced chemical plants) are too expensive, they simply won't happen. Our growth-based, globalized, consumption-oriented economy will require significant overhaul.

Though the prospect is daunting, this doesn't mean the renewable energy transition should not be attempted. As we wrote at the very beginning of this book (and will repeat again in part 3), the transition is both necessary and inevitable: maintaining our current fossil fuel–based energy system for much longer is simply not an option. However, this *does* mean that an all-renewable energy economy will have drawbacks as well as advantages (from our current perspective), and we should try to be realistic about both.

The advantages we will reap from an all-renewable energy economy will include the absence of financial and social costs associated with extracting, refining, transporting, and burning depleting fossil fuels—costs that will only increase as extractive industries have to drill deeper into lower-grade deposits; and the absence of the environmental externalities from burning those fuels— health and climate costs that would otherwise balloon to the trillions of dollars per year by midcentury. If we have fewer consumer products, they will likely be ones that are more durable and of higher quality. Because there will be tangible advantages to using energy when nature offers it, we are likely to feel more integrated into the rhythms of the natural world.

Perhaps it is helpful to maintain a long-term and philosophical view of our historical moment. Fossil fuels have enabled a temporary overshoot in human population levels and consumption patterns. Nevertheless, the planet is finite, and our energy use and population levels will inevitably be constrained—either voluntarily or otherwise. The renewable energy transition offers an opportunity to adapt to planetary limits more on our terms, preserving the best of what we have accomplished during our brief fling with nonrenewable energy sources. In a way, the renewable energy transition of the twenty-first century is a return of sorts. After all, for more than 99 percent of our species' history, we lived entirely on renewable sources of energy. Our challenge now is to learn to live within planetary limits while preserving the best of what we achieved during our brief, fossil-fueled binge of overconsumption.

PART III

Preparing for Our Renewable Future

CHAPTER 8

Energy and Justice

T HE ABILITY TO HARNESS ENERGY creates wealth and confers social power. With the advent of fossil fuels came a rush of wealth and power such as the world had never before seen. Naturally, humanitarians saw this as an opportunity to spread wealth and power around so as to lift all of humanity above drudgery, eliminate hunger, and even put an end to war. And to a large degree that opportunity has been seized: overall, child mortality rates are down, life expectancy is up, infectious diseases are on the decline, hunger has been reduced (even as population has dramatically grown), and mortality from violence has declined since the end of World War II.[1] Yet globally, the wealthy industrial nations have disproportionately benefited from the fossil fuel revolution while poorer nations have largely borne the costs. A similar disparity also exists within nations, both rich and poor ones. Further, the injustice of energy wealth versus energy poverty is increasingly magnified by climate impacts, which fall disproportionately upon energy-poor societies—both because of geographical happenstance and because they do not have the same level of resources to devote toward adaptation.

Today nearly two billion people live with the perks of cheap, abundant energy—plenty of affordable food, easy mobility, advanced health care, a surfeit of entertainment options, and more. But according to the United Nations 1.3 billion people do not have access to electricity, and 2.6 billion don't have access to clean cooking fuels and suffer a host of respiratory and cardiovascular diseases as a result.[2] Disadvantaged communities suffer the brunt of externalities from fossil fuel production (such as the environmental impacts of coal mining and combustion) but reap few if any of the benefits. From the coal miners of Appalachia, to the Native Americans whose land and health were devastated for uranium mining, to poor people in urban neighborhoods where coal-fired power plants and refineries are often sited, the story is depressingly familiar. And it continues, in the United States (with fracking sites in low income rural communities[3]), in China (where coal miners die by the thousands each year[4]), in Nigeria (where oil extraction ruins the land of poor farmers on Niger Delta while generating lavish incomes for well-connected businesspeople and politicians in Lagos and huge profits for Shell and Chevron[5]), and in dozens of other countries where fossil fuels are extracted.

One school of thought says that increased rates of energy flow through society actually create inequality. The recorded societies whose members have enjoyed the highest levels of economic equity are hunter-gatherer bands, in which individuals own little and share everything. These are also the societies with the lowest levels of energy consumption. As societies developed agriculture and then industry, full-time division of labor and the use of powered machines to generate wealth led reliably to the emergence of ruling and merchant classes, alongside large numbers of displaced landless peasants and economic refugees who were actually worse off, in many ways, than hunter-gatherers. Philosopher Ivan Illich epitomized this way of thinking in his 1974 book, *Energy and Equity,* wherein he wrote, "Below a threshold of per capita wattage, motors improve the conditions for social progress. Above this threshold, energy grows at the expense of equity. Further energy affluence then means decreased distribution of control over that energy."[6]

However, even societies with very high per capita levels of energy use have varying levels of economic inequality as measured by the Gini index,[7] which represents the income distribution within a nation (fig. 8.1). Typically, high-energy-flow societies achieve greater equity by taxing high incomes and inher-

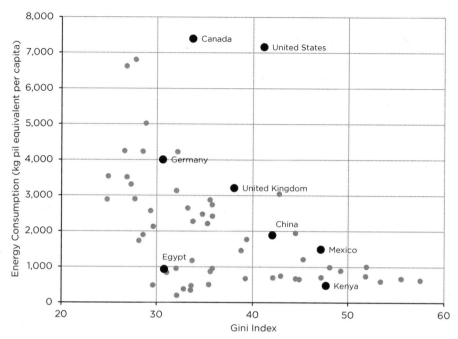

Figure 8.1. Per capita levels of energy use compared to levels of economic inequality in selected countries. The Gini index is a commonly used measure of inequity; its value for Egypt has not been updated since 2008, or for Kenya since 2005.
Source: World Bank, World Development Indicators, http://data.worldbank.org/data-catalog/world-development-indicators.

ited wealth, and through government redistributive programs (universal free health care, subsidized higher education, unemployment insurance, payments to retirees, and so on). Nevertheless, through trade such societies create poverty conditions elsewhere in the world (by encouraging ruinous and unnecessary indebtedness and by inducing regimes in poor nations to maintain inhumane labor conditions and inadequate environmental regulations), even though they manage to reduce and even mostly eliminate those conditions internally.

Now we arrive at a crossroads, where the wealth-generating energy sources of the past two centuries (fossil fuels) must give way to different energy sources. While the decades ahead may see declining per capita energy consumption in the wealthy industrialized world, the transition to renewable energy does not automatically herald a more egalitarian future. As everyone else adjusts to lower consumption levels, entrenched economic interests that benefited disproportionately during the fossil fuel era may seek to maintain their advantages, attempting to ensure that their slice of a diminishing pie is left untouched. It is also possible that nations, and wealthy communities within nations, will build robust, largely self-contained renewable energy systems while everyone else continues to depend upon increasingly dysfunctional and expensive electricity grids that are increasingly starved of fuel. In either case, current levels of economic inequality could persist or worsen.

Pursuing the renewable energy transition without equity in mind would likely doom the entire project. Unless the interests of people at all economic levels are taken into account and existing inequalities are reduced, the inevitable stresses accompanying this all-encompassing societal transformation could result in ever-deeper divisions both between and within nations. On the other hand, if everyone is drawn along into a visionary project that entails shared effort as well as shared gains, the result could be overwhelmingly beneficial for all of humanity. This is true not only for the renewable energy transition but also for our response to impacts of climate change that are by now unavoidable.

This chapter briefly traces some of the current economic fault lines between and within nations and surveys two of the frameworks that have been proposed to enhance equity or justice while simultaneously reducing carbon emissions and fossil fuel consumption.

Energy and Equity in the Least Industrialized Countries

Some nations (particularly in parts of sub-Saharan Africa and Central America) currently are desperately poor and need to increase their total and per capita energy consumption in order to achieve a living standard that is barely adequate. In these nations, access to food and water is problematic—not to mention housing, clothing, communication, and health care. Electricity, clean cookstoves, and refrigeration are rarely available. Often in these societies, the traditional village-based organization of society has more or less broken down as a result of war, rapid population growth, rapid urbanization, and global trade (organized under terms that tend to benefit the industrialized, wealthy nations), and there is little to replace it.

Conventional economic development aims to lift nations out of poverty by reproducing the process whereby currently wealthy nations obtained their wealth—that is, through resource extraction, manufacturing, urbanization, free market policies, debt, and trade, all based on ever-increasing consumption of fossil fuels. In this "development" process, most poor nations of the global South never seem to get (or are allowed to get) beyond the stage of resource extraction, and become trapped in an economic model in which a small elite class within the society gains control of whatever resources have trade value, thus monopolizing the country's wealth, most of which is exported to the global North. What is actually needed for these countries, in addition to political and economic reforms, is a pathway to secure food, water, education, electricity, transportation, communication, and health care that leapfrogs fossil fuels and fuel-dependent infrastructure, while respecting common ownership of resources and traditional community-based culture where it still persists.

In these currently least industrialized nations, the solution must include intermediate or appropriate technology—a tool set originally proposed by economist E. F. Schumacher in the 1970s.[8] These technologies are small-scale, decentralized, labor-intensive, energy-efficient, environmentally sound, and locally controlled. They include bicycle- and hand-powered water pumps (and other self-powered equipment), self-contained solar lamps and streetlights, and passive solar buildings that use local materials and respect traditional designs. To avoid expenses related to patents and licensing fees, appropriate technology can be (and often is) developed using open-source principles. For example, Open Source Ecology is developing, through open-source collaboration and experimentation, a "Global Village Construction Set"—fifty industrial machines, using modular parts that, in combination, it claims can build a small, sustainable civilization with modern comforts.[9] As energy consumption levels decline in currently highly industrialized nations, intermediate technology may increasingly serve human needs in these countries as well.

Further, instead of dealing with food shortages in poor countries by dumping food surpluses from wealthy industrial nations (which often results in the bankrupting of indigenous farmers who cannot economically compete with free food from aid agencies), the better long-term solution is to implement land reform to reopen large tracts of privatized agricultural land to small-holding farmers, then offer free education in low-energy agroecology and permaculture, again respecting traditional practices and cultural norms (the capacity to offer free education along these lines at scale currently does not exist, nor is it clear what agency could accomplish it, though small nonprofit organizations such as Ecology Action have made a start[10]). Proposals along these lines have been put forward for many years by Helena Norberg-Hodge and the International Society for Ecology and Culture, under the banner of "counterdevelopment."[11]

Energy and Equity in Rapidly Industrializing Nations

Other nations (e.g., China, India, and parts of Southeast Asia), now rapidly industrializing, are succeeding in building globally competitive manufacturing capacity with the use of cheap labor and cheap energy (typically from indigenous coal). From an international equity perspective, this might initially seem to be a path to progress, but it results in much higher global greenhouse gas emissions, as well as increased dependency on fossil fuel–reliant infrastructure. Crucially, these industrializing nations have also seen sharply rising domestic economic inequality,[12] along with profound health and environmental costs. A recent study found that outdoor air pollution contributed to 1.2 million premature deaths in China in 2010 alone.[13] In the context of what is needed for a successful and just energy transition, this conventional economic development pathway seems to hold little promise.

Indeed, for quickly industrializing countries, the energy transition would appear to require a profound directional shift. Typically, only two or three decades ago these nations consisted mostly of subsistence farmers. Policy makers have followed the example of the countries that were first to industrialize, with the clear goal of building a consumption-oriented middle class. They have systematically discouraged subsistence agriculture in favor of industrialized agriculture, built fossil fuel–dependent urban infrastructure, and promoted ever-expanding manufacturing for global trade using coal as a primary energy source. These nations are well on the road to achieving their goals. But since, in the process, they are generating greater income inequality as well as crippling

Pollution in China. (Credit: testing, via Shutterstock.)

levels of environmental pollution and unsustainable dependence on fossil fuels, those goals, as well as their methods for achieving them, require complete
revision.

One way or another, the trend toward urbanization in rapidly industrializing countries will taper and perhaps even reverse itself. Expanding cities require
more capacity to transport people and goods. In a world with less liquid transport fuel, cities will need to prioritize clustered, compact, mixed-used development patterns and nonmotorized and electrified transportation infrastructure.[14] Additionally, developing countries may do well to find ways to make life
in existing smaller cities and towns and villages more rewarding and economically attractive: despite their energy efficiency, megacities like Shanghai and
Mexico City have real problems with the overconcentration of pollution, traffic, and resource use (e.g., water). The flexibility of hundreds of medium-sized
cities with a diversity of resources, leaders, and policies may be more suited to

meeting the shifting challenges of the low-energy future. In addition, subsistence agriculture must be supported rather than economically discouraged.

At the highest level, the goal of these nations must not be to emulate the growth pathways of wealthy industrial nations of the twentieth century, because the economies of those latter nations depended upon unsustainable fossil fuel use, and the per capita energy consumption of already industrialized nations needs to be curbed to meet climate and equity objectives (and, as we have seen, probably will shrink in any case). A better goal would be to achieve a level of per capita energy consumption that is sustainable over the long term using renewable energy sources (what that level is, exactly, is presently unknown, though it is almost certainly much lower than the current per capita level of consumption in older industrialized nations), and to do so in a way that promotes economic equity. Ultimately, the per capita energy consumption of the already industrialized, and the rapidly industrializing, nations must converge on that sustainable and equitable mean.

Rapidly industrializing nations will benefit from a rapid reduction in pollution levels, which are currently resulting in millions of early deaths annually. A shift in focus from economic growth to improvement in quality of life could yield social, political, and health benefits.

To say that this will be a complicated and difficult transformation in priorities and goals is surely an understatement. To name just one quandary: currently China, the largest of the rapidly industrializing nations, is the world leader in the manufacturing of photovoltaic panels, which are produced using coal for high-temperature industrial heat and also for electricity. How will China wean itself from coal while actually increasing its production of photovoltaic panels?

Energy and Equity in Highly Industrialized Countries

For the United States, Canada, Australia, the countries of western Europe, and a few other nations, a significant transitional problem will consist of reducing per capita energy consumption equitably. Clearly, energy consumption correlates to a large extent with income and wealth; thus equity would be served by, for example, a tax on carbon-based energy consumption that primarily targets high-rate users (most of whom would be wealthy and thus capable of paying it).

However, this would by itself be insufficient. In the highly industrialized nations, more efficient ways of living are often out of the reach of lower-income people. Poorer members of industrial societies typically rent rather than own their homes, so energy-efficiency upgrades are the responsibility of landlords, many of whom are unwilling to make such investments. If they do own their homes, low-income families still may be unable to afford the costs of home insulation, double-pane windows, solar hot water heaters, air-source heat pumps, photovoltaic panels, and so forth. Government programs will be needed to help low-income homeowners make such upgrades, and government regulations requiring—and low-interest loans or other assistance to help—landlords to invest in them will also be needed.

Low-income people increasingly can't afford to live in desirable central locations in cities, with walkable neighborhoods and good public transit. Pushed out to far-flung suburbs, what they may save in rent can be rapidly consumed by the cost of owning and maintaining a car and driving long distances to meet daily needs. Cities and suburbs will need to be redesigned so that all people have good alternatives to private car ownership, with a focus on mixed-use and clustered development. Transportation priorities will need to shift profoundly, with new road building coming to a halt and investment shifting to infrastructure for public transit, bicycling, and walking, particularly the revitalization of electrified public transit between and within communities.

Since the economic crisis of 2008–2009 there has been an upwelling of political and economic discussion about inequality within industrialized nations. Many people are aware that wages have stagnated, partly as a result of globalization and mechanization.[15] Proposals merely to stimulate more consumption, manufacturing, and trade—the twentieth century solutions to stagnation and inequality—will not work during the renewable energy transition, at least not in the same way. The parts of the economy that will require stimulus—and that will accommodate an increase in consumption, manufacturing, and trade—are those related to renewable energy (solar panels, wind turbines, energy storage, grid upgrades) and energy efficiency (building retrofits, rail revitalization, public transit). Some other parts of the economy may need to shrink significantly as investment capital and energy are directed to these key transitional sectors. Policy will need to be crafted to make sure the burden of these shifts does not fall too heavily on workers in shrinking industries, by providing skills and training that will be relevant in the renewable future.

As globalization stalls and retreats as a result of constraints and trends outlined in chapter 4, it will be important to rebuild local economies—local manufacturing, investment, and food systems. This in itself will offer opportunities for increasing equity and justice, through the formation and promotion of local cooperative institutions (co-ops and credit unions), and through the devolution of a great deal of political organization and decision making. Localization efforts can create jobs that pay living wages and help individuals within the community develop critical skills that directly benefit themselves and their neighbors.[16] They can build resilience in communities that face a future filled with economic and environmental challenges. And such efforts can focus on the inclusion of groups that have historically been disadvantaged.

Finally, the forms of ownership adopted for new renewable energy systems will likely go a long way toward determining the degrees of economic equity or inequity in the renewable future. Centralized ownership through for-profit corporations will tilt the playing field toward continued accumulation of wealth in fewer hands; distributed generation and ownership of generating capacity and grid-related infrastructure by communities, through cooperative, nonprofit financing and revenue-sharing models, will result in more equity. Equity and justice will not be automatic outcomes of relocalization. They will require intentional, organized effort and struggle.

Decentralized energy democracy could be a significant driver of equity. Rooftop solar (whether on single- or multifamily buildings) frees electricity consumers from monopoly utility companies charging monthly bills. The current utility system also distributes pollution unequally, with most going to the poor. Our current energy system (a centralized system dominated by fossil fuels) is inherently regressive. A decentralized grid, with decentralized ownership models, has the potential to be inherently progressive. Further, while distributed generation, distributed storage, microgrids, and community choice aggregation all serve to create a more equitable power infrastructure, they can also provide technical advantages, such as resource diversity and system resilience.

Policy Frameworks for Enhancing Justice While Cutting Carbon

The problem of reversing historic energy-related economic injustice during the renewable transition must be addressed not just within nations but also between and among nations. Two policy frameworks aim to deal with climate change and international inequality at the same time.

Greenhouse Development Rights

This framework, developed and modeled by Paul Baer and Tom Athanasiou of EcoEquity, along with Sivan Kartha and Eric Kemp-Benedict of the Stockholm Environment Institute, aims to show how the costs of rapid climate stabilization can be shared fairly among countries.[17] Greenhouse development rights (GDRs) represent national "fair shares" in the costs of an emergency global climate mobilization, based on two factors: responsibility for contributing to climate change (fig. 8.2) and capacity to act. Responsibility and capacity are defined relative to a "development threshold" that exempts the poorest nations from national obligations.

This threshold is set above the global poverty line (about $16 per person per day, in purchasing power parity terms), reflecting a level of welfare beyond the most basic needs but well below levels of consumption in the industrialized world. People with higher incomes assume a larger proportion of the costs of curbing emissions, as well as the costs of providing low-emissions pathways for development of those still living below the threshold. These obligations are taken to belong to everyone living above the development threshold, regardless of whether they live in countries that are rich or poor overall.

GDRs constitute a framework by which fair shares of emissions rights can be calculated and negotiated, and against which existing climate treaties and strategies can be evaluated and compared.

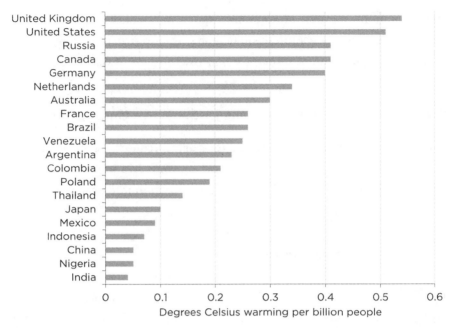

Figure 8.2. Selected countries' historic per capita contribution to climate change.
Source: H. Damon Matthews et al., "National Contributions to Observed Global Warming," *Environmental Research Letters* 9, no. 1 (January 15, 2014), doi:10.1088/1748 -9326/9/1/014010.

Common Wealth Trusts

This is a more general framework, pioneered by Peter Barnes of the Tomales Bay Institute, that can be applied not just to the atmosphere and climate but also to other natural resources that represent common wealth—that is, wealth belonging to everyone equally—and that therefore could and should be organized legally to be protected for all, and to benefit all equally.[18]

Organizing common wealth so that markets respect its co-inheritors and co-beneficiaries requires the creation of legal shells, in the form of common wealth trusts that are legally accountable to future generations. These trusts would have authority to limit usage, charge for use, and pay per capita dividends. The shells are necessary to enable managers of common wealth to bargain with profit-seeking enterprises that would seek to use the resources in question. Fiduciary responsibility assures that the managers of common wealth act first and foremost on behalf of future generations.

Outwardly, the shells would be not-for-profit corporations with state charters, self-governance, and (within the United States, at least) legal personhood. Inwardly, the managers of these not-for-profit corporations would be required to protect their assets for future generations and to share current income (if any) for the common good.

The trust form of organization need not be applied to all forms of common wealth. However, at the very minimum, it could be applied to ecosystems (such as the global atmosphere and climate) that are approaching irreversible tipping points. In these cases, trusts would ratchet aggregate usage steadily downward, limiting human impact to scientifically determined levels. Prices would then be determined by the forces of demand and (now limited) supply. For sources of common wealth that are global, including the atmosphere and oceans, global negotiations would be necessary for harmonizing these limits.

Barnes recommends distributing the proceeds from the trusts to all citizens equally. In the United States, organized common wealth could, by his calculation, over time generate enough income to pay a dividend of up to $5000 per person per year.

There are precedents for this approach. Large-scale public funds (also known as sovereign wealth funds) already exist in some countries (Norway[19]) and several American states—notably the Alaska Permanent Fund, which derives income from oil and gas extraction within the state, most of which is distributed equally to all Alaskan citizens.[20] However, other states use revenues from such funds on social services (primarily public education). In Texas, revenues from the Texas Permanent School Fund—which owns and manages millions of acres of land in perpetuity—support public schools in every county and city both through direct transfers and bond guarantees.[21]

This raises the question as to whether it would be more effective or even more equitable in the long run to distribute revenues from a common wealth trust as universal basic income (as suggested by Barnes) or on social or environmental spending. There is little evidence that a basic income alone will provide incentive for citizens to adopt more environmentally benign practices (it has not done so in Alaska). Moreover, for individuals at higher wealth and income distributions, a universal basic income could serve to further increase consumption, if prices of goods are relatively stable. On the other hand, if the cost of goods increases due to increasing energy costs, a guaranteed income could serve to ensure that those increased costs would be at least partially compensated for. In this case dollar spending would increase, though not necessarily material consumption. Accordingly, using at least some revenues from common wealth trusts for other public and environmental purposes may prove to be more effective at increasing environmental sustainability and reducing social inequality, at least in the short term.

An example more relevant to this report is California's cap-and-trade program, which collects dividends on auctioned allowances to emit greenhouse gases. The California emissions cap declines each year so that about 3 percent fewer allowances are allocated per annum, ensuring that these allowances will become more valuable as time passes.[22] This approach incentivizes large emitters to invest in emission reductions sooner rather than later. For 2015, annual revenue from the auctions is expected to exceed $2.5 billion. Funds generated are dedicated into two broad categories: 25 percent goes to ameliorate impacts on low-income groups affected by the policy or by climate change, the remainder is directed to statewide capital investments in renewable energy, conservation, public transportation, and research to transition California away from climate-destabilizing economic activity. Early evidence suggests that the program is succeeding in encouraging low-carbon infrastructure investments, both from the market response to the rising cost of emission allowances and from state investments in low-carbon infrastructure.[23] However, cap-and-trade programs have been critiqued by economic justice advocates who say that pollution already disproportionately affects low-income communities and communities of color, and creating a carbon-trading system only exacerbates those trends.[24]

* * *

Clearly there are potential roadblocks to these frameworks, and to other ideas and proposals in this chapter. Rich countries, and wealthy individuals and legal entities within countries, are unlikely to be willing to part with their economic advantages, especially during a time when they'll be experiencing an economic pinch anyway as a result of the withdrawal of fossil fuels from global manufacturing and transport systems. Only social action through organized movements could force them to do so.

In any case, the equity and justice questions won't go away. From the perspective of global elites, something must be done to level the playing field and take everyone's interests into account (whether through an overarching global framework or through piecemeal national and regional efforts), or those who feel excluded will disrupt efforts toward an orderly energy transition. From the perspective of those with far lower levels of power and wealth, there is no reason to support efforts to reduce fossil fuel consumption if those efforts only preserve or exacerbate economic inequality. To succeed, climate and energy policies need to be grounded in a universally shared and ethically based agreement that all human beings, regardless of income, gender, or ethnicity, have both the right to a safe and stable environment, and the responsibility to act in such a way as to sustain and protect it.

CHAPTER 9

What Government Can Do

THIS BOOK'S SURVEYS of renewable energy price trends, opportunities for renewable energy deployment, and challenges to that deployment, lead us to conclude that market mechanisms by themselves will be insufficient to drive the renewable energy transition at the speed required to outrun climate change and fossil fuel depletion. Government policy will be required to direct sufficient capital toward building renewable energy capacity, to manage the build-out of energy storage and necessary grid upgrades, to manage the evolution of industries (transportation, agriculture, manufacturing, mining) that currently rely on nonelectricity uses of fossil fuels, and to provide efficiency incentives and mandates to ease the burden of a likely decline in overall energy availability during the transition.

Current government policy, in the United States and globally, is simply not up to these tasks. To mention just one example, current research into renewable energy, energy storage, and energy transmission accounts for only about 1 percent of government research and development spending in the world's wealthy industrial countries (fig. 9.1).[1,2] Far more is spent on weapons research. But clearly the problems of climate change and fossil fuel depletion constitute at least as great a threat to world peace and security as does military aggression (indeed, the Pentagon has described climate change as a "threat multiplier"[3]). Without sufficient capital spending, that threat will become an almost certain source of unprecedented human misery and environmental disruption.[4]

What is needed is a sense of the overall goals, challenges, and opportunities of the energy transition, and a phased approach that takes into account both the necessity and costs of the transition, while also distributing those costs in such a way that crucial sectors (such as agriculture) are not seriously compromised.

We see five primary areas in which better policies are needed; in most but not all cases, either pilot policies are in place in at least some countries or communities, or potentially useful policy frameworks have been suggested by other authors and organizations:

1. Support for an overall switch from fossil fuels to renewable energy
2. Support for research and development of ways to use renewables to power more industrial processes and transport
3. Conservation of fossil fuels for essential purposes
4. Support for energy conservation in general—efficiency and curtailment
5. Better greenhouse gas (GHG) accounting

Support for an Overall Switch from Fossil Fuels to Renewable Energy

Considerable policy research has already been devoted to this goal, especially for the electricity sector, with four primary proposals to advance state and national renewable electricity targets[5] gaining the most interest; some have seen limited practical implementation.

Feed-in tariffs (also known as standard-offer contracts) are discussed in chapter 3 of this book, where we noted Germany's relative success with this strategy. Other nations, including Australia, Canada, France, India, Israel, Spain, the United Kingdom, and the United States have also used feed-in tariffs.[6] As we have already pointed out, while prices of wind and solar electricity are falling, a rapid transition still requires subsidies. The key to success seems to

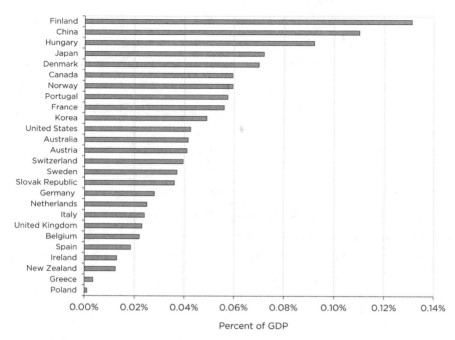

Figure 9.1. Selected countries' public investments in renewable energy research and development.
Source: Bill Gates, "We Need Energy Miracles," Gatesnotes, the Blog of Bill Gates, June 25, 2014, http://www.gatesnotes.com/Energy/Energy-Miracles.

be long-term commitment and continual adjustment of tariffs to reflect market conditions. There are two obvious pitfalls: too high a tariff—which encourages a flood of speculative money that in turn encourages the building of projects, some of which are ill sited or ill conceived; or too low a tariff—which fails to attract sufficient investment capital.

An even bigger pitfall of feed-in tariffs is improper design. Spain, for example, created an inflexible tariff tied to its tax system, rather than to rates that were adjusted annually or semiannually as in the German system (see chapter 3).[7]

Renewable energy mandates (renewable portfolio standards, or RPSs) are requirements that a certain percentage of electricity be produced from renewable sources. These have been implemented in Australia, China, Europe, Japan, and the United States among other nations.[8] Within the United States, twenty-nine states plus the District of Columbia, including California, Colorado, Kansas,

Michigan, Nevada, New York, North Carolina, Ohio, and Texas, have all established RPS policies.[9] One pitfall seems to be that real or imagined hikes in electricity prices attributable to the shift to wind and solar power can be used by interest groups to persuade lawmakers to abandon this approach. Thus West Virginia has recently removed its RPS, Ohio has frozen its target for two years, Kansas has made its targets voluntary (although this was after achieving its target five years ahead of schedule), and conservative legislators in North Carolina have twice (unsuccessfully) attempted to overturn its 12.5 percent target.[10] Again, to succeed, this strategy must have long-term commitment from policy makers, who must recognize that electricity prices may eventually be affected and who must prepare their constituents for cost increases, both psychologically and through policies that reduce impacts on low-income households and particularly vulnerable industries. In fact, electricity price increases attributable to RPSs have so far been negligible (a detailed analysis from the National Renewable Energy Laboratory showed that "over the 2010–2012 period, average estimated incremental RPS compliance costs in the United States were equivalent to 0.9% of retail electricity rates"[11]). Policy makers must provide incentives and mandates for utilities to build the infrastructure necessary for higher rates of renewable energy penetration in the grid, so that electricity prices do not more steeply increase as the transition proceeds to the point where renewable electricity represents roughly 40 percent or more of the overall mix.

Carbon taxes or cap-and-auction policies that dedicate at least some of their revenues toward renewables would generate investment capital to build renewable energy production capacity and could also be used for raising capital for nonelectricity segments of the renewable energy transition, as well as for energy efficiency and conservation. We have already discussed California's cap-and-trade policy, which aims to funnel revenues toward statewide capital investments in renewable energy production capacity, energy conservation, public transportation, and energy research.[12]

Carbon taxes, cap-and-trade, cap-and-auction, and cap-and-dividend policies have been widely discussed in the literature on climate change policy. The consensus in that literature seems to be that the renewable energy transition cannot proceed far or fast enough without at least one of these kinds of policies in place in all industrialized and fast-industrializing nations (though they would have differing implications for equity).

There are three questions that need resolution: Would a tax (or fee) approach be as effective as a simple cap? What should be the level of the emissions cap? And what should be done with the proceeds?

Some see carbon caps as preferable to fees or taxes. If properly set, properly adjusted annually, and properly enforced, caps would prevent carbon fuels from being extracted and burned, and would do so at a planned and regulated pace commensurate with the need for climate protection. Promoters of carbon taxes or fees aim to increase the price of carbon; they would then either use the revenues to fund the renewable energy transition or redistribute them to offset the rising cost of energy—which would promote equity. However, while putting a price on carbon may discourage extraction and consumption of fossil fuels, it does not definitively specify the quantities of carbon-based fuels that will in fact be burned (or not burned). If the fees or taxes are redistributed, since the wealthy would be paying proportionally more in taxes (because they consume more carbon fuels), the net result will be an increase in income to the poor.

Proponents of a carbon tax counter that there is a level of taxation that would actually reduce fossil fuel consumption because affordability has been cut significantly. That level has never been tested. However, if taxation raises the prices of everything (because everything has an energy component), then the poor would only increase material consumption if their income transfers were greater than the price increases.

With a carbon cap, companies would be paying more for fuels (for manufacturing and transport) and would raise the prices of their products to recoup their higher costs. People with lower incomes would thus have to pay more for their limited purchases; however, it is at least possible that the share of the carbon revenue they received would initially more than compensate for the higher prices of goods, since high-carbon users (the rich) would be paying proportionally more in taxes. On a net basis, then, the poor might still tend to benefit.

A cap-and-share policy (i.e., a policy in which revenues from emissions-permit auctions are rebated to low-income persons) would ensure emissions reductions while promoting equity. If low-income families gain, their spending would rise—and since there are far more poor people than rich ones, aggregate spending would likely increase. But with carbon capped, that spending would have to go toward meeting human needs in low-carbon ways. Low-carbon enterprises would constitute the growing parts of the economy, and new employment would likely be generated. However, some of the capital raised from auctions might still be needed to fund much of the transition directly, in terms of capacity build-out, research and development, manufacturing process redesign, and grid upgrades.

Tradable energy quotas (TEQs), though little discussed outside the United Kingdom, represent another distinct emissions and energy trading scheme that deserves consideration. TEQs constitute an electronic energy rationing system designed to be implemented at the national scale.[13] Every adult would be given an equal free entitlement of TEQ units each week, while other energy users (government and industry) would bid for their units at a weekly auction. Anyone using less than their entitlement of units could sell the surplus; anyone needing more could buy them. All trading would take place at a single national price, which would rise and fall in line with demand. Buying and selling would take place electronically.

When buying carbon-based energy, such as gasoline, units corresponding to the amount of energy purchased would be deducted from the individual's TEQ account, in addition to the monetary payment. The total number of units available in the country would be set out in a TEQ budget, with the size of the budget declining each year. Since the national TEQ price would be determined by national demand, it would be transparently in everyone's interest to help reduce energy demand, encouraging a national sense of common purpose.

TEQs could be revenue neutral, or yield a financial surplus to be invested in the energy transition. In either case, the net effect would be to incentivize an overall reduction in carbon-based energy usage, thus incentivizing noncarbon energy sources.

In addition to these four policy mechanisms, nations, regions, and local governments are increasingly adopting 100 percent renewable energy targets to accelerate the transition and make it more efficient. According to research by the Renewables 100 Policy Institute, more than a hundred government entities worldwide have committed to, achieved, or surpassed a 100 percent renewable energy target in at least one sector (though virtually all of these focus just on electricity).[14]

Support for Research and Development of Ways to Use Renewables to Power More Industrial Processes and Transport

Carbon taxes, cap-and-auction policies, or tradable energy quotas could be used in part to support research and development for the expansion of renewable energy into industrial processes and transportation. However, more money could likely be freed up for this purpose, and more quickly, simply by redirecting existing research and development funding. There would be grumbling from military contractors, which are the primary current beneficiaries of government research grants; but military-related research into renewable energy is already on the increase, and at least some of that research is likely to benefit society as a whole.[15] Government regulations and mandates could also be used to encourage key industries to undertake such research—for example, by requiring cement producers to reduce carbon emissions from their activities by a certain percentage each year.

In the United States, much transport policy is crafted at the state, county, and even municipal levels. This may create complications for a national shift away from road building and toward rail-based transport options. However, it also opens opportunities: in the absence of forward-thinking national policy, states and communities can change priorities on their own. For example, the state of California has often led the nation in automobile emissions (and other product) standards, with those higher standards quickly being adopted by the rest of the country simply because manufacturers do not want to make products to differing standards.[16]

Conservation of Fossil Fuels for Essential Purposes

As discussed in chapter 6, society's remaining economically viable fossil fuels will be crucially important to the energy transition, since we will need to use them to fuel the building of our renewable future even as we phase them out to avert catastrophic climate change. Very little policy thinking has so far addressed this conundrum. However, even a moment's thought suggests that any solution would have to entail altering the current purely market-based allocation scheme for fossil fuels. Perhaps industries involved in the direct manufacturing of renewable energy technologies could be partially or entirely shielded from carbon taxes or other policy devices intended to discourage fossil fuel consumption. However, in order to compensate, other fossil fuel users would have to ratchet down their consumption further and faster than would otherwise be the case. We will develop this discussion further in chapter 11.

Support for Energy Conservation in General—Efficiency and Curtailment

Some of the policies already surveyed (carbon taxes, TEQs) would incentivize reduction in energy usage. However, there are many other policies that can be pursued at all levels of government toward this goal.

At the federal level, funding for research and development could help in increasing the efficiency of products and processes throughout the industrial system, in every building, and in every appliance. Government could also simply mandate efficiency improvements (as the United States has recently done with regard to automobile fuel efficiency). Research on higher efficiency standards actually suggests they increase innovation overall and that prices are lower than expected.[17] Trade policies must shift from favoring globalization toward import substitution, using subsidies and tariffs to promote local production for local consumption wherever practical, so as to reduce reliance on transport fuels. And agricultural policy must shift away from support for fossil fuel–intensive farming toward smaller-scale, more ecologically oriented production, promoting local food systems and soil conservation.

At the state and local level, governments will probably have the greatest impact through policies and investments that impact land use and transportation, two deeply connected issues that together determine how much energy will be needed to move people and goods.[18]

For land use, local leaders can use development programs, zoning codes, and building codes to reduce communities' need for energy outright. Mixed-use nodal neighborhoods make it easier for people to walk and bike to meet their daily needs. Multifamily housing saves energy because it is easier to heat a large, single building than multiple small buildings—and it creates the density necessary for economically viable mixed-use and public transit. Building codes can include simple requirements like better insulation and daylighting to further reduce energy needs, or more stringent requirements that get closer to zero-energy and even net-energy-positive buildings. In addition, "cradle-to-cradle" or "circular economy" principles can be adopted into regulations to discourage waste and maximize recycling and reuse.

For transportation, it is well within the power of local leaders to stop building roads (which facilitate the expansion of the most energy-intensive of our transport options) and to instead invest more in energy efficient modes of transportation like public transit, bicycling, and walking. Typically, municipal leaders need citizen encouragement in such efforts. Portland, Oregon, pioneered just such an approach for American cities when it reappropriated federal highway funds to build the country's first major urban light rail line in the 1980s, an act initiated by citizen activists. Decades later, thanks to a mix of activism, business community support, and state and local government policymaking, Portland is now famous for its people-centered transportation infrastructure—and its consistently declining per capita energy use and GHG emissions reflect this.[19]

Better Greenhouse Gas Accounting

Without reliable information about how and where GHG emissions are produced, it will be difficult to make effective policies to reduce those emissions. Unfortunately, current national GHG emissions accounting methods tend to be somewhat misleading.

Traditional production-based accounting measures emissions at production sources, and thus significantly underestimates the GHGs emitted in international trade.[20] For example, when China burns coal to manufacture smart phones destined for the United States, the emissions are attributed entirely to

China. This method needs to be supplemented with consumption-based accounting, which accounts for GHG emissions embodied in manufactured goods purchased by end users.[21,22] Consumption-based GHG accounting yields a fairer and more accurate view of the world in which wealthier countries are responsible for a greater share of GHG emissions.

International aviation and shipping are the only GHG-emitting sectors not covered by the international accounting (this remains true following the Paris COP21 agreement).[23] GHG emissions in these sectors, which together amount to more than 6 percent of total global emissions, do not show up on the emissions accounts of any nation.[24,25]

* * *

Beyond the realm of legislation and regulation, policy makers must recognize and accept their necessary role in helping reshape public attitudes about energy. Climate change and the renewable energy transition should not merely constitute one small subset of a blinding variety of media obsessions ranging from local murders to the affairs of pop stars. Instead, the energy transition needs to become the organizing context within which we see and understand everything else that is happening in the world. It needs to be the next great global project, akin to mobilization efforts in the United States for World War II—when Americans were asked to conserve, recycle, and grow their own food.

We all must come to share the common understanding that climate change and our response to it constitute a wartime level of emergency, and that we all must cooperate toward a common goal. This shift in mass awareness is unlikely to occur unless and until opinion leaders and policy makers themselves fully understand what is at stake. And that will require pressure from citizens and nongovernmental organizations—as well as the business sector, which is profoundly vulnerable to climate disruption.

World War II poster. (Credit: National Archives of the United States.)

CHAPTER 10

What We the People Can Do

SOUND NATIONAL AND INTERNATIONAL CLIMATE policies are crucial: without them, it will be impossible to organize a transition away from fossil fuels and toward renewable energy that is orderly enough to maintain industrial civilization, while speedy enough to avert catastrophic ecosystem collapse. However, world leaders have been working on hammering out effective climate policies for nearly a quarter of a century, and during that time greenhouse gas emissions have continued to increase. And the impacts of climate change are becoming ever more incontrovertible and perilous. Clearly, individuals, households, communities, and nongovernmental organizations cannot merely stand by and hope that political leaders somehow find the wherewithal at the last moment (if it is not already too late) to halt our descent into climate chaos. We must put all possible pressure on those leaders to take politically difficult decisions to severely limit carbon emissions.

That will require collective action on a scale that has yet to be seen. The massive transformations in energy systems, government, and the economy that we have described are exceedingly unlikely to occur absent struggle and social action. Powerful interests invested in the extractive economy will not give up their advantages willingly. As Frederick Douglass eloquently said, "Power concedes nothing without a demand. It never did and it never will."

At the same time, we must also show that we as citizens are ready for climate policies by proactively reducing our reliance on fossil fuels and cutting our greenhouse gas emissions. In the process, we can road-test behaviors and technologies that are needed on a broader scale. Fortunately, many people, communities, and organizations have already started doing this, but more are needed.

Individuals and Households

Tackling the energy transition, climate change, and energy inequality will require collective action and policy. So the most important thing we can do as individuals is to support equitable solutions to climate change, and support local democracy and engagement in local decisions about energy.

Nevertheless, our personal actions and choices also reverberate through our communities and can back our words with the authority of personal experience. Start by doing what you can to reduce your use of energy in general, and especially of fossil fuels. That requires developing awareness and changing habits. How much energy do you use? Where and how? Find out by doing a personal and household energy audit. Don't just look at your electricity consumption (though that's essential); also examine your gasoline and natural gas usage. Then make a plan, using a footprint calculator.[1]

Most likely, it will be a long-term plan that will be implemented in stages. In some cases, it will require investment—perhaps in superinsulating your house; perhaps in exchanging your current automobile for a small electric car; or perhaps in installing an air-source heat pump, a solar water heater, a solar cooker, a front-loading washing machine, a clothesline, and insulated cookware.[2] If you rent your home, some of these purchases may be less feasible unless you can come to an agreement with your landlord to share costs and savings. If you live in an area where you have no choice but to drive virtually everywhere, you might consider moving to a more compact, mixed-use neighborhood that doesn't require you to spend so much energy just to meet your daily needs—and move into a smaller home with lower heating, cooling, and maintenance needs while you're at it.

Other parts of your plan will be devoted to changing habits: using public transit or bicycling more (if that infrastructure is available), reducing the frequency of shopping trips (and buying less overall), shortening showers, and turning off appliances when not in use. You might also consider what you eat: some food choices (such as beef) involve far more embodied water and energy than others (such as whole grains). Reducing carbon emissions means reducing both operational and embodied energy consumption—not just having more efficient machines, but fewer of them, and replacing them less frequently. It means eating lower on the food chain, wearing clothes longer before discarding them, and repairing goods that break wherever possible, rather than replacing them.

Support the expansion of renewable energy in your community by signing up to purchase clean energy through your utility. Not all utility companies offer this option, but many do.[3] Buy or lease solar panels for your home or business. Aggregate with your neighbors to find ways to get good deals on solar panels, or support community choice aggregation or "go solar" via shared/community solar programs where those are legal. Where those things aren't allowed, get involved politically and make them legal!

Support relocalization efforts in your town by buying local wherever possible. That means making purchases at locally owned shops, and banking at locally owned banks and credit unions. But it also means looking for and preferentially buying locally grown food and locally made products. If you have pension funds or other investments, it is also possible to invest locally to support local economic development.[4]

Overall, get involved with local efforts to advance the transition to renewable energy. In over forty-five countries and over 2000 cities and towns around the world, Transition Initiatives inspire individuals, families, and neighborhoods to adopt strategies to reduce fossil fuel consumption, localize economies, and produce more renewable energy.[5]

Communities

Often the most important steps toward catalyzing the energy transition within communities takes the form of efforts to build public awareness about climate and energy. Such efforts can be driven by elected officials, but are more likely to gain traction if led or co-led by citizen groups.

There is also a growing movement to push cities, towns, and counties to make commitments to be 100 percent renewably powered (sometimes this concerns electricity only, sometimes the commitment is more broadly conceived). These are exciting new citizen-led efforts that you can join or start. For example, in Sonoma County, California, a group called the Center for Climate Protection[6] has helped create a local power provider, designed a pilot program for water and energy conservation, and persuaded leaders of all the cities in the region to sign on to stringent greenhouse gas reduction targets. Many other communities have aggregated this way in the states where it is legal. Even in states where it's not, many cities and communities have at least committed to go 100 percent renewable (or provide some degree of rooftop solar) on public buildings.

Community Choice Aggregation (CCA), is a system adopted into law in the states of Massachusetts, New York, Ohio, California, New Jersey, Rhode Island, and Illinois that enables cities and counties to aggregate the buying power of individual customers within a defined jurisdiction so as to secure alternative energy supply contracts on a community-wide basis. Households that don't wish to participate can opt out. As of 2014, CCAs serve nearly 5 percent of Americans in over 1300 communities.[7] Many CCAs purchase and sell a higher percentage of renewable energy than their conventional utility competitors; some also offer a "green power" option at a slightly higher rate, enabling customers to purchase 100 percent renewable energy. In California, local governments have been using CCAs as a tool to achieve higher greenhouse gas reductions and renewable electricity procurement targets than state requirements mandate or than competing independently owned utilities. But this has not always been true in other states, where cost reduction is the main goal. When renewable energy is cheaper, they procure more of it; but when it's not, CCAs in other states often revert to conventional fuel procurement. CCAs also help achieve equity by promoting local control over energy sources.

As previously noted, in the United States state and local governments can have considerable control over the built environment of communities, which influences how people use energy. Land use and transportation plans determine for decades whether people will be able to walk, bike, and take public transit to meet their daily needs, or will have no choice but to drive (and likely own) a car. Zoning policies and building regulations within communities can either encourage or discourage cohousing developments and other manifestations of the sharing economy, as well as natural buildings and zero-energy buildings. Typically, municipal leaders need citizen encouragement in such efforts. Often regulations change as the result of pioneering efforts by individuals and small groups willing to organize their neighbors, meet (and argue) extensively with local officials, and patiently sit through many city council meetings to keep the political pressure on.

With "Go Local" programs thriving in hundreds of cities across the country, localism is growing into a community effort across America. Perhaps the most important thrust of relocalization efforts (and the easiest to organize) is the push for relocalized food systems. The United States Department of Agriculture currently lists 8144 farmers markets in its National Farmers Market Directory, up from 5000 in 2008.[8] Indeed, local food is one of the fastest-growing segments of American agriculture. Further steps communities can take to promote local economic resilience include analyses to determine the proportion of food, energy, goods, and services that come from local sources.

Efforts to relocalize economic activity usually start with citizen groups. In Santa Rosa, California, a citizen-organized Go Local campaign has resulted in a downtown storefront that is home to Share Exchange—perhaps best described as a localist mini-mall, hosting a "Made Local" marketplace, a "share space" co-working center, and a cooperative business incubator. Signs on Santa Rosa windows and lampposts advise residents to "Shop Local," "Bank Local," "Eat Local," and "Compost Local." Menus at an upscale restaurant at the center of town proclaim, "We feature organic food from local farmers."

Ultimately, localization means changing economic development goals. This can be an involved, detailed, and contentious process. The Sustainable

Economies Law Center in Oakland, California, is one resource; it offers legal guidance in building community resilience and grassroots economic empowerment, highlighting policy recommendations for sharable cities.[9]

Climate and Environmental Groups, and Their Funders

When considering the role of climate and environmental groups, perhaps it is useful to start by listing some important things already accomplished by climate and energy nongovernmental organizations:

- They have changed the conversation about fossil fuels and climate change through a divestment campaign, which persuades investors to sell stocks or bonds issued by oil, coal, and gas companies. This largely symbolic campaign casts fossil fuel companies in roughly the same light as South Africa's apartheid regime, which was targeted by similar divestment campaigns in the 1980s.
- They have proposed policies to further the goal of climate justice—that is, to make sure that the impacts of climate change and the costs of climate adaptation do not fall disproportionately on poorer nations, and to help poor nations leapfrog fossil fuel–based development pathways and build renewables-based economies capable of providing a sustainable, globally equitable per capita level of consumption.
- They have documented fossil fuel health and environmental impacts and exposed the public relations lies of fossil fuel industries in denying climate change, and in denying the culpability of their own products.
- They have campaigned for energy efficiency and proposed and studied specific ways of reducing energy consumption in many sectors of society.
- They have begun to prepare society for impacts of climate change by studying the factors that make communities more resilient in the face of disruption.

These are important contributions, and much more along these lines is still needed. However, there are some other tasks that have so far received less emphasis from environmental organizations:

- Give citizens a realistic sense of the size and scope of the energy transition, and help prepare society for an effort and a shift as huge as the Industrial Revolution.
- Identify key uses of fossil fuels that will be hardest to substitute (aviation fuel, for example), and argue for workarounds (such as rail) or for the managed shutdown of those uses.
- Explore how the transition could provide satisfying livelihoods and support thriving localized, steady-state, circular economies.
- In addition to resisting the dominance of fossil fuels, engage with communities to create persuasive models of how people can live and thrive with much reduced reliance on fossil fuels.

The philanthropic sector inevitably exerts a very large influence over the priorities of nonprofit organizations that it funds. Funders should increasingly support the following:

- Efforts to educate and inspire citizens about the energy transition
- Projects that involve development of new economic models that enable people to live with less energy, but in ways that bring greater life satisfaction
- Replicable models of community development that include taking charge of local energy production and reducing fossil fuel demand across many sectors

Funders could also help the nonprofit community view the energy transition as a systemic transformation, one that only *begins* with shutting down coal power plants.

CHAPTER 11

What We Learned

THE AUTHORS BEGAN THIS BOOK project with some general understanding of the likely energy transition constraints and opportunities; nevertheless, researching and writing *Our Renewable Future* has been a journey of discovery. Along the way, we identified not only technical issues requiring more attention, but also important implications for advocacy and policy. What follows is a very short summary, tailored mostly to the United States, of what we've learned.

We Really Need a Plan; No, Lots of Them

Germany has arguably accomplished more toward the transition than any other nation, largely because it had a plan—the *Energiewende*, which we discussed in chapter 3. This plan targets a 60 percent reduction in all fossil fuel use (not just in the electricity sector) by 2050, achieving a 50 percent cut in overall energy use through efficiency in power generation, buildings, and transport. It is not a perfect plan, in that it really should aim higher than 60 percent. But it is certainly better than nothing, and the effort is off to a good start. The United States does not have an equivalent official plan. Without it, we are at a significant disadvantage.

What would a plan do? It would identify the low-hanging fruit, show how resources need to be allocated, and identify needed policies. We would of course need to revise the plan frequently as we gained practical experience (as Germany is doing).

What follows are some components of a possible plan, based on work already done by many researchers in the United States and elsewhere; far more detail (with timelines, cost schedules, and policies) would be required for a fleshed-out version. We've grouped tasks into levels of difficulty; work would need to commence right away on tasks at all levels, but for planning purposes it is useful to know what can be achieved relatively quickly and cheaply, and what will take long, expensive, sustained effort.

Level One: The "Easy" Stuff

Nearly all energy transition analysts agree that the easiest way to kick-start the transition would be to replace coal with solar and wind power for electricity generation. That would require building lots of panels and turbines while regulating coal out of existence. Distributed generation and storage (rooftop solar panels with home- or office-scale battery packs) will help. Replacing natural gas will be harder, because gas-fired "peaking" plants are often used to buffer the intermittency of industrial-scale wind and solar inputs to the grid (see "Level Two").

As we've noted repeatedly, electricity accounts for less than a quarter of all final energy used in the United States (see fig. 3.1). What about the rest of the energy we depend on? Since solar, wind, hydro, and geothermal produce electricity, it makes sense to electrify as much of our energy usage as we can. For example, we could heat and cool most buildings with electric air-source heat pumps (replacing natural gas- or oil-fueled furnaces). We could also begin replacing all our gas cooking stoves with electric stoves.

Transportation represents a large swath of energy consumption, and personal automobiles account for most of that. We could reduce oil consumption substantially if we all drove electric cars (replacing 250 million gasoline-fueled automobiles will take time and money but will eventually result in energy and financial savings). But promoting walking, bicycling, and public transit will take much less time and investment, and be far more sustainable in the long term.

Buildings will require substantial retrofitting for energy efficiency (this will again take time and investment but will offer still more opportunities for savings). Building codes should be strengthened to require net-zero energy or near-net-zero-energy performance for new construction. Zoning codes and development policy should encourage infill development, multifamily buildings, and clustered mixed-use development. More energy-efficient appliances will also help.

The food system is a big energy consumer, with fossil fuels used in the manufacturing of fertilizers, in food processing, and transportation. We could reduce a lot of that fuel consumption by increasing the market share of organic (i.e., not using synthetic fertilizers, herbicides, and pesticides) local foods. While we're at it, we could begin sequestering enormous amounts of atmospheric carbon in topsoil by promoting farming and land management practices that build soil rather than depleting it.

If we got a good start in all these areas, we could achieve at least a 40 percent reduction in carbon emissions in ten to twenty years.

Level Two: The Harder Stuff

As we've seen, solar and wind technologies have a drawback: they provide energy intermittently. When they become dominant within our overall energy mix, we will have to accommodate that intermittency in various ways. We'll need substantial amounts of grid-level energy storage as well as a major grid overhaul to get the electricity sector to 80 percent renewables (thereby replacing natural gas in electricity generation). We'll also need to start timing our energy usage to coincide with the availability of sunlight and wind energy. That in itself will present both technological and behavioral hurdles.

Electric cars aside, the transport sector will require longer-term and sometimes more expensive substitutions. We could reduce our need for cars (which require a lot of energy for their manufacture and decommissioning) by densifying our cities and suburbs and reorienting them to public transit, bicycling, and walking. We could electrify all motorized human transport by building more electrified public transit and intercity passenger rail links. Heavy trucks could run on fuel cells, but it would be better to minimize trucking by expanding freight rail. Transport by ship could employ sails to increase fuel efficiency (this is already being done on a tiny scale), but relocalization or deglobalization of manufacturing would be a necessary co-strategy to reduce the need for shipping.

Much of the manufacturing sector already runs on electricity, but some critical aspects don't—and many of these will offer significant challenges. Many raw materials for manufacturing processes either *are* fossil fuels (feedstocks for plastics and other petrochemical-based materials, including lubricants, paints, dyes, pharmaceuticals, etc.) or currently require fossil fuels for mining or transformation (e.g., most metals). Considerable effort will be needed to replace fossil fuel–based materials and to recycle nonrenewable materials more completely, significantly reducing the need for mining.

If we did all these things, while also building far, far more solar panels and wind turbines, we could achieve roughly an 80 percent reduction in emissions compared to our current level.

Level Three: The Really Hard Stuff

Doing away with the last 20 percent of our current fossil fuel consumption is going to take still more time, research, and investment—as well as much more behavioral adaptation. Just one example: we currently use enormous amounts of concrete for all kinds of construction activities, and concrete requires cement. As we've seen, cement making needs high heat, which could theoretically be supplied by sunlight, electricity, or hydrogen—but that will entail a nearly complete redesign of the process.

While with Level One we began a shift in food systems by promoting lo-cal organic food, driving carbon emissions down further will require finishing that job by making *all* food production organic, and requiring *all* agriculture to build topsoil rather than depleting it. Eliminating all fossil fuels in food systems will also entail a substantial redesign of those systems to minimize processing, packaging, and transport.

The communications sector—which uses mining and high heat processes for the production of phones, computers, servers, wires, photo-optic cables, cell towers, and more—presents some really knotty problems. The only good long-term solution in this sector is to make devices that are built to last a very long time and then to repair them and fully recycle and remanufacture them when absolutely needed. The Internet could be maintained via the kinds of low-tech, asynchronous networks now being pioneered in poor nations, using relatively little power.[1]

Back in the transport sector: we've already made shipping more efficient with sails, but doing away with petroleum altogether will require costly substi-tutes (fuel cells or biofuels). One way or another, global trade will almost inevi-tably shrink. There is no good drop-in substitute for aviation fuels; we may have to write off aviation as anything but a specialty transport mode. Planes running on hydrogen or biofuels are an expensive possibility, as are dirigibles filled with (nonrenewable) helium, any of which could help us maintain vestiges of air travel. Paving and repairing roads without oil-based asphalt is possible, though it will require an almost complete redesign of processes and equipment.

The good news is that if we do all these things, we can get to *beyond* zero carbon emissions; that is, with sequestration of carbon in soils and forests, we could actually reduce atmospheric carbon with each passing year.

Plans will look different in each country, so each country (and each state) needs its own.

Scale Is the Biggest Challenge

When we performed the thought exercise of starting with a blank page and de-signing a renewable energy system that (1) has minimal environmental impacts,

(2) is reliable, and (3) is affordable, we found this could easily be done in several different ways—as long as relatively modest amounts of energy were needed. Once current U.S. scales of energy production and usage were assumed we found we had to sacrifice the environment (because of the vast tracts of land needed for siting wind turbines and solar panels), reliability (because of the intermittency of solar and wind), or affordability (because of the need for storage or capacity redundancy). Power is a secondary hurdle: ships and airplanes require energy-dense fuels because they are maneuvering such enormous weights. Renewable energy resources can supply the needed power, but once again scale is the issue: building and operating a few hydrogen-powered airplanes for specialized purposes would certainly be technically feasible, but operating fleets of thousands of commercial planes using hydrogen fuel is daunting from both technical and economic perspectives.

It's Not All About Solar and Wind

These two energy resources have been the subjects of most of the discussion surrounding the renewable energy transition. Prices are falling, rates of installation are high, and there is a large potential for further growth (fig. 11.1). However, as we have pointed out repeatedly, the inherent intermittency of these energy sources will pose increasing challenges as percentage levels of penetration into overall energy markets increase. Other renewable energy sources—hydropower, geothermal, and biomass—can more readily supply controllable base load power, but they have much less opportunity for growth.

Hopes for high levels of wind and solar are therefore largely driven by the assumption that industrial societies can and should maintain very high levels of energy use. Once again, the challenge is scale: if energy usage in the United States could be scaled back significantly (70 to 90 percent) then a reliable all-renewable energy regime—based more upon hydro, geothermal, and biomass, but with solar and wind used in situations where intermittency is not a problem—becomes much easier to envision and cheaper to engineer.

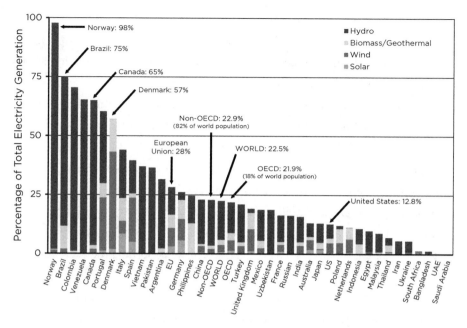

Figure 11.1. Percentage of electricity generated by renewables in selected countries, 2014. *Source*: J. David Hughes, Global Sustainability Research, Inc. (data from BP Statistical Review, 2015).

We Must Begin Preadapting to Having Less Energy

As we saw in chapter 6, it is unclear how much energy will be available to society at the end of the transition: there are many variables (including rates of investment and the capabilities of renewable energy technology without fossil fuels to back them up and to power their manufacture, at least in the early stages). Nevertheless, given all the challenges involved, it would be prudent to assume that people in wealthy industrialized countries will have less energy (even taking into account efficiencies in power generation and energy usage) than they would otherwise have, assuming a continuation of historic growth trends.

This conclusion is hard to avoid when considering the speed and scale of reduction in emissions actually required to avert climate catastrophe. As climate scientist Kevin Anderson points out in an upcoming *Nature Geoscience* paper: "According to the IPCC's Synthesis Report, no more than 1,000 billion metric tons (1,000 Gt) of CO_2 can be emitted between 2011 and 2100 for a 66% chance (or better) of remaining below 2°C of warming (over preindustrial times). . . . However, between 2011 and 2014 CO_2 emissions from energy production alone amounted to about 140 Gt of CO_2." Subtracting realistic emissions budgets for deforestation and cement production, "the remaining budget for energy-only emissions over the period 2015–2100, for a 'likely' chance of staying below 2°C, is about 650 Gt of CO_2."[2]

That 650 gigatons of carbon amounts to less than nineteen years of continued business-as-usual emissions from global fossil energy use. The notion that the world could make a complete transition to alternative energy sources, using only that nineteen-year fossil energy budget, and without significant reduction in overall energy use, might be characterized as optimism on a scale that stretches credulity.

The "how much will we have?" question reflects an understandable concern to maintain current levels of comfort and convenience as we switch energy sources. But in this regard it is good to keep ecological footprint analysis in mind.

According to the Global Footprint Network's *Living Planet Report 2014*, the amount of productive land and sea available to each person on Earth in order to live in a way that's ecologically sustainable is 1.67 global hectares.[3] The current per capita ecological footprint in the United States is 6.8 global hectares (if the entire world population lived at this footprint it would require four planet Earths) (fig. 11.2). Asking whether renewable energy could enable Americans to maintain their current lifestyle is therefore equivalent to asking whether renewable energy can keep us living unsustainably. The clear answer is: only temporarily, if at all—so why attempt the impossible? We should aim for a sustainable level of energy and material consumption, which on average is significantly lower than at present.

Efforts to preadapt to shrinking energy supplies have understandably gotten a lot less attention from activists than campaigns to leave fossil fuels in the ground, or to promote renewable energy projects. But if we don't give equal thought to this bundle of problems, we will eventually be caught short, and there will be significant economic and political fallout.

Ecologically sustainable equitable distribution
1.67 gha per person

At European per capita ecological footprint
4.5 gha per person

At U.S. per capita ecological footprint
6.8 gha per person

Figure 11.2. How many Earths does it take? Productive global hectares (gha) per capita. *Source*: Global Footprint Network.

So what should we do to prepare for energy reduction? Look to California: its economy has grown for the past several decades while its per capita electricity demand has not. The state encouraged cooperation between research institutions, manufacturers, utilities, and regulators to figure out how to keep demand from growing by changing the way electricity is used.[4] This is not a complete solution (California's population has grown during this period, so its total electricity consumption has also grown; we do not have a good example of absolute reduction in aggregate energy use). Nevertheless this may be one of the top success stories in the energy transition so far, rivaling that of Germany's *Energiewende*. It should be copied in every state and country.

Consumerism Is a Problem, Not a Solution

Current policy makers see increased buying and discarding of industrial products as a solution to the problem of stagnating economies. With nearly 70 percent of the U.S. economy tied to consumer spending,[5] it is easy to see why consumption is encouraged. Historically, the form of social and economic order known as *consumerism* largely emerged as a response to industrial overproduction—one of the causes of the Great Depression—which in turn resulted from an abundant availability of cheap fossil energy.[6] Before—but especially after—the Depression and World War II, the advertising and consumer credit industries grew dramatically as means of stoking product purchases, and politicians of all persuasions joined the chorus, urging citizens to think of themselves as "consumers," and to take their new job description to heart.

If the transition to renewable energy implies a reduction in overall energy availability, if mobility is diminished, and if many industrial processes involving high heat and the use of fossil fuels as feedstocks become more expensive or are curtailed, then conservation must assume a much higher priority than consumption in the dawning post-fossil-fuel era. If it becomes more difficult and costly to produce and distribute goods such as clothing, computers, and phones, then people will have to use these manufactured goods longer, and repurpose, remanufacture, and recycle them wherever possible. Rather than a consumer economy, this will be a *conserver* economy.

The switch from one set of priorities and incentives (consumerism) to the other (conservation) implies not just a major change in American culture but also a vast shift both in the economy and in government policy, with implications for nearly every industry. If this shift is to occur with a minimum of stress, it should be thought out ahead of time and guided with policy. We see little evidence of such planning currently, and it is not clear what governmental body would have the authority and capacity to undertake it. Nor do we yet see a culture shift powerful and broad-based enough to propel policy change.

The renewable economy will likely be slower and more local. Economic growth may reverse itself as per capita consumption shrinks; if we are to avert a financial crash (and perhaps a revolution as well), we may need a different economic organizing principle. In her recent book on climate change, *This Changes Everything*, Naomi Klein asks whether capitalism can be preserved in the era of climate change; while it probably can (capitalism needs profit more than growth), that may not be a good idea because, in the absence of overall growth, profits for some will have to come at a cost to everyone else. And this is exactly what we have been seeing in the years since the financial crash of 2008 (fig. 11.3).

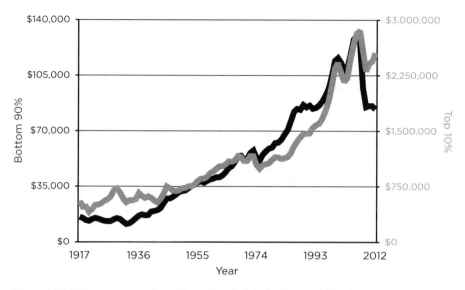

Figure 11.3. U.S. average real wealth per family (chained 2010 dollars).
Source: E. Saez, University of California Berkeley, http://eml.berkeley.edu/~saez/#income.

The idea of a conserver economy has been around at least since the 1970s, and both the European degrowth movement[7] and the leaders of the relatively new discipline of ecological economics[8] have given it a lot of thought. Their insights deserve to be at the center of energy transition discussions.

Population Growth Makes Everything Harder

A discussion of population might seem off-topic for this book. But if energy and materials (which represent embodied energy) are likely to be more scarce in the decades ahead of us, population growth will mean even less consumption per capita. And global population is indeed growing: on a net basis (births minus deaths) we are currently adding 82 million humans to the rolls each year,[9] a larger number than at any time in the past, even if the percentage *rate* of growth is slowing.

Population growth of the past century was enabled by factors—many of which trace back to the availability of abundant, cheap energy and the abundant, cheap food that it enabled—that may be reaching a point of diminishing returns. Policy makers face the decision now of whether to humanely reduce population by promoting family planning and by public persuasion, by raising the educational level of poor women around the world and giving women full control over their reproductive rights, or by letting nature deal with overpopulation in unnecessarily brutal ways. For detailed recommendations regarding population matters, consult population organizations such as Population Institute[10] and Population Media Center.[11] Population is a climate and energy issue.

Fossil Fuels Are Too Valuable to Allocate Solely by the Market

Our analysis suggests that industrial societies will need to keep using fossil fuels for some applications until the very final stages of the energy transition—and possibly beyond, for nonenergy purposes. Crucially, we will need to use fossil fuels (for the time being, anyway) for industrial processes and transportation needed to build and install renewable energy systems.[12] We will also need to continue using fossil fuels in agriculture, manufacturing, and general transportation, until robust renewable energy–based technologies are available. This implies several problems.

As the best of our remaining fossil fuels are depleted, society will by necessity be extracting and burning ever-lower-grade and/or harder-to-get coal, oil, and natural gas. We see this trend already far advanced in the petroleum industry, where virtually all new production prospects involve tight oil, tar sands, ultraheavy oil, deepwater oil, or Arctic oil—all of which entail high production costs and high environmental risk as compared to conventional oil found and produced during the twentieth century—and refining what are sometimes heavier, dirtier fuels (in the case of tar sands) creates ever more co-pollutants that have a disproportionate health impact and burden on low-income communities. The fact that the fossil fuel industry will require ever-increasing levels of investment per unit of energy yielded has a gloomy implication for the energy transition: much of society's available capital will have to be directed toward the deteriorating fossil fuel sector to maintain current services, just as much more capital is also needed to fund the build-out of renewables. Seemingly the only way to avoid this trap would be to push the energy transition as quickly as possible, so that we aren't stuck two or three decades from now still dependent on fossil fuels that, by then, will be requiring so much investment to find and extract that society may not be able to afford the transition project.

But there is also a problem with accelerating the transition too much. Since we use fossil fuels to build renewables, speeding up the transition could mean an overall increase in emissions—unless we reduce other current uses of fossil fuels (if the pace of end-use electrification exceeds the pace of renewable energy electricity production growth, then this could also lead to higher emissions). In other words, we may have to deprive some sectors of the economy of fossil fuels before adequate renewable substitutes are available, in order to fuel the transition without increasing overall greenhouse gas emissions. This would translate to a reduction in overall energy consumption and in the economic benefits of energy use (though money saved from conservation and efficiency would hopefully reduce the impact), and this would have to be done without producing a regressive impact on already vulnerable and economically disadvantaged communities.

We may be entering a period of fossil fuel triage. Rather than allocating fossil fuels simply on a market basis (those who pay for them get them), it may be fairer, especially to lower-income citizens, to find ways to allocate fuels based on the strategic importance of the societal sectors that depend on them, and on the relative ease and timeliness of transitioning those sectors to renewable substitutes. Agriculture, for example, might be deemed the highest priority for continued fossil fuel allocations, with commercial air travel assuming a far lower priority. Perhaps we need not just a price on carbon, but different prices for different uses. We see very little discussion of this prospect in the current energy policy literature. Further, few governments even currently acknowledge the need for a carbon budget. The political center of gravity, particularly in the United States, will have to shift significantly before decision makers can publicly acknowledge the need for fossil fuel triage.

As fossil fuels grow more costly to extract, there may be ever-greater temptation to use our available energy and investment capital merely to maintain existing consumption patterns, and to put off the effort that the transition implies. If we do that, we will eventually reap the worst of all possible outcomes—climate chaos, a gutted economy, and no continuing wherewithal to build a bridge to a renewable energy future.

Everything Is Connected

Throughout the energy transition, great attention will have to be given to the interdependent linkages and supply chains connecting various sectors (communications, mining, and transport knit together most of what we do in industrial societies). Some links in supply chains will be hard to substitute, and chains can be brittle: a problem with even one link can imperil the entire chain. This is the modern manifestation of the old nursery rhyme, "for the want of a nail, the kingdom was lost."

Consider, for example, the materials required to manufacture and operate a wind turbine. Figure 11.4 shows the various components, each with its own manufacturing sector somewhere in the world.

Planning will need to take such interdependencies into account. As every ecologist knows, *you can't do just one thing.*[13]

This Really Does Change Everything

Energy transitions change societies from bottom to top, and from inside out. From a public relations standpoint, it may be helpful to give politicians or the general public the impression that life will go on as before while we unplug coal power plants and plug in solar panels, but the reality will probably be quite different. During historic energy transitions, economies and political systems underwent profound metamorphoses. The agricultural revolution and the fossil-fueled Industrial Revolution constituted societal watersheds. We are on the cusp of a transformation every bit as decisive.

*   *   *

We end this book as we began it, by restating our firm conviction that the transition from fossil fuels to renewable energy is necessary and inevitable. But, as has been shown, this transition will not be an automatic or simple process. There are many potential roadblocks, some of which arise from simple inertia: companies—indeed, whole societies—will invest in fundamental changes to their ways of doing business only when they have to, and most are quite

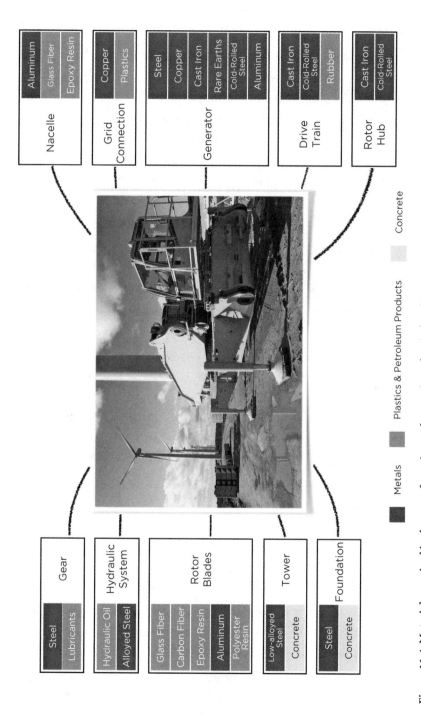

Figure 11.4. Materials required in the manufacturing and operation of a wind turbine.
Source: Till Zimmermann, Max Rehberger, and Stefan Gößling-Reisemann, "Material Flows Resulting from Large Scale Deployment of Wind Energy in Germany," *Resources* 2, no. 3 (2013): 303–34, doi:10.3390/resources2030303.

comfortable with their current fossil-fuel-dependent processes, supply chains, and of course sunk costs.

Studies claiming that a transition to renewable energy will be easy and cost-free may allay fears and thus help speed the transition. However, sweeping actual difficulties under the carpet also delays confronting them. Our society needs to start now to address the problems of energy demand adaptation, of balancing intermittency in energy supply from solar and wind (or, preferably, finding ways to use variable energy sources at the times of their greatest abundance), and of energy substitution in thousands of industrial processes. Those are big jobs, and ignoring them won't make them go away.

If many of the unknowns in the renewable energy transition imply roadblocks and speed bumps, some could turn out to be opportunities, and we cheerfully acknowledge that many conundrums may be much more easily solved than currently appears likely. For example, it is conceivable that new technical advances could result in a zero-carbon cement that is cheaper to make than the current carbon-intensive variety.[14] But that is extremely unlikely to happen until serious attention is given to the problem.

At the end of the renewable energy transition, if it is successful, we will achieve savings in ongoing energy expenditures needed for each increment of economic production, and we may be rewarded with a quality of life that is acceptable and perhaps preferable over our current one (even though, for most Americans, material consumption will be scaled back from its current unsustainable level). We will get a much more stable climate than would otherwise be the case, along with greatly reduced health and environmental impacts from energy production activities. However, the conversion to 100 percent renewable energy will not by itself solve other environmental issues facing us—including deforestation, land degradation, and species extinctions, among others.

A point we have raised repeatedly is that possibly the most challenging aspect of this transition is its implication for economic growth: whereas the cheap, abundant energy of fossil fuels enabled the development of a consumption-oriented growth economy, renewable energy will likely be unable to sustain such an economy. Rather than planning for continued, unending expansion, policy makers must begin to imagine what a functional postgrowth economy could look like. Among other things, the planned obsolescence of manufactured goods must end, in favor of far more durable products that can be reused, repaired, remanufactured, or recycled indefinitely.

To us, given factors currently visible and the unknowns arrayed ahead, it seems wise to channel society's efforts toward no-regrets strategies (that is, actions to save energy that make sense in view of a range of possible futures)—efforts that shift expectations, emphasizing quality of life over consumption; and efforts that result in increased community resilience. Even though it may be impossible to fully envision the end result of the renewable energy transition, we believe that it is essential for society to seek to gain a sense of its scope and general direction. That is why we have written this book.

One way or another, our descendants a few decades from now will inhabit an all-renewable world (or nearly so), and it will be a world that works differently, in many significant ways, from the world we know today. It could be a better world in which to live, or it could be much worse, depending on the decisions we make in the next decade or two. Right now society is putting off even the most obvious and pressing of those decisions (starting with a mandatory global cap on carbon emissions). Successive waves of problems and requirements for decision will follow. Failing to see those next waves from a distance only makes the worse possibilities for our renewable future more likely. We hope that this exploratory effort shines a light into the future implications of the renewable energy transition, so that we can start now to see and understand the territory, consider our options, and act intelligently.

Notes

Introduction

1. United Nations, Framework Convention on Climate Change, *Adoption of the Paris Agreement*, FCCC/CP/2015/L.9 (December 12, 2015), http://unfccc.int/resource /docs/2015/cop21/eng/l09.pdf.
2. Steven Mohr et al., "Projection of World Fossil Fuels by Country," *Fuel* 141 (2015): 120–135, doi:10.1016/j.fuel.2014.10.030.
3. Andrew Nikiforuk, *The Energy of Slaves: Oil and the New Servitude* (Vancouver, BC: Greystone Books, 2012).
4. Mikael Höök et al., "Hydrocarbon liquefaction: Viability as a peak oil mitigation strategy," *Philosophical Transactions of the Royal Society A* 372, no. 2006 (2014): 20120319, doi:10.1098/rsta.2012.0319. David Murphy, Charles Hall, and Bobby Powers, "New perspectives on the energy return on (energy) investment (EROI) of corn ethanol," *Environment, development and sustainability* 13, no. 1 (2011): 179–202.
5. See chapter 5 in Michael Carolan, *Cheaponomics: The High Cost of Low Prices* (New York: Routledge, 2014).

Chapter 1

1. Charles Hall and John Day. "Revisiting the Limits to Growth after Peak Oil." *American Scientist* 97, no. 3 (2009): 230–237. Michael Dale, Susan Krumdieck, and Pat Bodger, "Net Energy Yield from Production of Conventional Oil." *Energy Policy* 39, no. 11 (2011): 7095–7102.

2. David Murphy, "The Implications of the Declining Energy Return on Investment of Oil Production," *Philosophical Transactions of the Royal Society A* 372, no. 2006 (2014): 20130126, doi:10.1098/rsta.2013.0126.

3. Michael Carbajales-Dale, Susan Krumdieck, and Pat Bodger, "Global Energy Modelling—A Biophysical Approach (GEMBA) Part 2: Methodology," *Ecological Economics* 73 (2012): 158–167, doi:10.1016/j.ecolecon.2015.06.010. Michael Dale and Sally Benson, "Energy Balance of the Global Photovoltaic (PV) Industry: Is the PV Industry a Net Electricity Producer?," *Environmental Science and Technology* 47, no. 7 (2013): 3482–3489, doi:10.1021/es3038824.

4. Jessica Lambert et al., "Energy, EROI and Quality of Life," *Energy Policy* 64 (2014): 153–167, doi:10.1016/j.enpol.2013.07.001.

5. Charles Hall, Stephen Balogh, and David Murphy, "What Is the Minimum EROI That a Sustainable Society Must Have?" *Energies* 2, no. 1 (2009): 25–47.

6. Michael Dale and Sally M. Benson, "Energy Balance of the Global Photovoltaic (PV) Industry: Is the PV Industry a Net Electricity Producer?," *Environmental Science and Technology* 47, no. 7 (2013): 3482–3489.

7. Richard Heinberg, "The Brief, Tragic Reign of Consumerism—and the Birth of a Happy Alternative," *Post Carbon Institute, Articles + Blog*, April 14, 2014, http://www.postcarbon.org/the-brief-tragic-reign-of-consumerism-and-the-birth-of-a-happy-alternative/.

8. U.S. Department of Agriculture Economic Research Service, "Food Availability (Per Capita) Data System," accessed September 30, 2015, http://www.ers.usda.gov/data-products/food-availability-(per-capita)-data-system/summary-findings.aspx.

9. Charles Hall and Kent Klitgaard, *Energy and the Wealth of Nations: Understanding the Biophysical Economy* (New York: Springer, 2012), 73.

10. Sassan Saatchi et al., "Benchmark Map of Forest Carbon Stocks in Tropical Regions across Three Continents," *Proceedings of the National Academy of Sciences* 108, no. 24 (2011): 9899–9904. Saatchi et al. report between 100 to 700 (oven dry) Mg/ha above-ground biomass in tropical forests, which converts to 40 to 283 Mg/acre (1 ha = 2.47 acre); the Biomass Energy Centre reports a net calorific value of 19 MJ per metric ton of wood (http://www.biomassenergycentre.org.uk/portal/page?_pageid=75,20041&_dad=portal). Thus 40,000 to 283,000 kg/acre × 19 MJ / kg = ~750,000 to 5,377,000 MJ/acre.

11. Harald Sverdrup, Kristin Vala Ragnarsdottir, and Deniz Koca, "An Assessment of Metal Supply Sustainability as an Input to Policy: Security of Supply Extraction Rates, Stocks-in-Use, Recycling, and Risk of Scarcity," *Journal of Cleaner Production*, June 2015, doi:10.1016/j.jclepro.2015.06.085.

12. National Research Council, *Coal: Research and Development to Support National Policy, National Academy of Sciences* (Washington, DC: National Academies Press, 2007). Emily Grubert, "Reserve Reporting in the United States Coal Industry," *Energy Policy* 44 (2012): 174–84, doi:10.1016/j.enpol.2012.01.035.

13. International Energy Agency, *Key World Energy Statistics* (Paris: OECD/IEA, 2014).
14. Patrick Moriarty and Damon Honnery, "What Is the Global Potential for Renewable Energy?" *Renewable and Sustainable Energy Reviews* 16 (2012): 244–52, doi:10.1016/j.rser.2011.07.151.
15. U.S. Environmental Protection Agency, "Energy Projects and Candidate Landfills," last modified March 13, 2015, http://www3.epa.gov/lmop/projects-candidates/.
16. U.S. Energy Information Administration, "Coal Transportation Rates to the Electric Power Sector," Table 1 and Table 3c, accessed October 1, 2015, http://www.eia.gov/coal/transportationrates/.
17. U.S. Energy Information Administration, "How Much Electricity Is Lost in Transmission and Distribution in the United States?," accessed October 1, 2015, http://www.eia.gov/tools/faqs/faq.cfm?id=105&t=3.

Chapter 2

1. International Energy Agency, *World Energy Outlook 2015* (Paris: OECD/IEA, 2015), http://www.worldenergyoutlook.org/weo2015/.
2. Ezra Krendel, "9.1 Sources of Energy, Muscle-Generated Power," in *Marks' Standard Handbook for Mechanical Engineers*, 11th ed., ed. Eugene Avallone, Theodore Baumeister III, and Ali Sadegh (New York: McGraw Hill, 2007).
3. International Energy Agency, "World: Balance (2012)," accessed October 1, 2015, http://www.iea.org/sankey/#?c=World&s=Balance.
4. International Energy Agency, "World: Balance (2012)." Input 5083 (Mtoe)/losses 2784 (Mtoe) = 54% loss rate.

Chapter 3

1. International Energy Agency, "World: Balance (2012)." International Energy Agency, "United States: Final Consumption (2012)," accessed October 1, 2015, http://www.iea.org/sankey/#?c=United States&s=Final consumption.
2. BP, "Data Workbook—Statistical Review 2015," accessed October 2, 2015, http://www.bp.com/en/global/corporate/energy-economics/statistical-review-of-world-energy/downloads.html.
3. BP, "Data Workbook—Statistical Review 2015."
4. Chris Mooney, "Here's How Much Faster Wind and Solar Are Growing Than Fossil Fuels," *Washington Post*, March 9, 2015.
5. Vishal Shah, Jerimiah Booream-Phelps, and Susie Min, "2014 Outlook: Let the Second Gold Rush Begin," Deutsche Bank, January 6, 2014, https://www.deutschebank.nl/nl/docs/Solar_-_2014_Outlook_Let_the_Second_Gold_Rush_Begin.pdf.

6. Deborah Lawrence, "Investment in Solar Stocks Crushed Big Oil," *Energy Policy Forum*, November 4, 2014, http://energypolicyforum.com/2014/11/04/investment-in-solar-stocks-crushed-big-oil/.

7. James. Martinson, "The True Benefits of Wind Power," *Newsweek*, April 21, 2015, http://www.newsweek.com/true-benefits-wind-power-323595.

8. U.S. Energy Information Administration, "Table 6.7.B. Capacity Factors for Utility Scale Generators Not Primarily Using Fossil Fuels, January 2013–July 2015," accessed October 2, 2015, http://www.eia.gov/todayinenergy/detail.cfm?id=11991.

9. Michael Dale and Sally M. Benson, "Energy Balance of the Global Photovoltaic (PV) Industry: Is the PV Industry a Net Electricity Producer?" (see chap. 1, n. 6).

10. Mark Schwartz, "Stanford Scientists Calculate the Carbon Footprint of Grid-Scale Battery Technologies," *Stanford Report*, March 5, 2013, http://news.stanford.edu/news/2013/march/store-electric-grid-030513.html.

11. U.S. Energy Information Administration, "Pumped Storage Provides Grid Reliability Even with Net Generation Loss," *Today In Energy*, July, 8, 2013, http://www.eia.gov/todayinenergy/detail.cfm?id=11991.

12. Charles Barnhart and Sally Benson, "On the Importance of Reducing the Energetic and Material Demands of Electrical Energy Storage," *Energy & Environmental Science* 6, no. 4 (2013): 1083–92, doi:10.1039/C3EE24040A.

13. Tom Murphy, "Pump Up the Storage," *Do the Math*, November 15, 2011, accessed October 2, 2015, http://physics.ucsd.edu/do-the-math/2011/11/pump-up-the-storage/.

14. David Biello, "Inside the Solar-Hydrogen House: No More Power Bills—Ever," *Scientific American*, June 19, 2008, http://www.scientificamerican.com/article/hydrogen-house. See also Shannon Page and Susan Krumdieck, "System-Level Energy Efficiency Is the Greatest Barrier to Development of the Hydrogen Economy," *Energy Policy* 37, no. 9 (2009): 3325–35, doi:10.1016/j.enpol.2008.11.009.

15. Matthew Pellow et al., "Hydrogen or Batteries for Grid Storage? A Net Energy Analysis," *Energy and Environmental Science* 8 (2015): 1938–52, doi:10.1039/C4EE04041D.

16. Matthew Pellow et al., "Hydrogen or Batteries for Grid Storage?"

17. Alice Friedemann, "Making the Most Energy Dense Battery from the Palette of the Periodic Table," *Energy Skeptic*, April 15, 2015, http://energyskeptic.com/2015/making-the-most-energy-dense-battery-from-the-palette-of-the-periodic-table/.

18. Mark Schwartz, "Stanford Scientists Calculate the Carbon Footprint of Grid-Scale Battery Technologies."

19. Mark Schwartz, "Stanford Scientists Calculate the Carbon Footprint of Grid-Scale Battery Technologies."

20. Charles Barnhart, Michael Dale, Adam Brandt, and Sally Benson, "The Energetic Implications of Curtailing versus Storing Solar-and Wind-Generated Electricity," *Energy & Environmental Science* 6, no. 10 (2013): 2804–10.

21. Kris De Decker, "Off-Grid: How Sustainable Is Stored Sunlight," *Low-Tech Magazine*, accessed October 1, 2015, http://www.lowtechmagazine.com/2015/05/sustainability-off-grid-solar-power.html.

22. Shalke Cloete, "The Fundamental Limitations of Renewable Energy," *Energy Collective*, September 6, 2013, http://theenergycollective.com/schalk-cloete/257351/fundamental-limitations-renewable-energy.

23. Mark Jacobson et al. "Low-Cost Solution to the Grid Reliability Problem with 100% Penetration of Intermittent Wind, Water, and Solar for All Purposes," *Proceedings of the National Academy of Sciences USA* 112, no. 49 (December 8, 2015): 15060–65, doi:10.1073/pnas.1510028112.

24. International Energy Agency, Energy Technology Systems Analysis Programme and International Renewable Energy Agency, *Thermal Energy Storage: Technology Brief*, January 2013, https://www.irena.org/DocumentDownloads/Publications/IRENA-ETSAP%20Tech%20Brief%20E17%20Thermal%20Energy%20Storage.pdf.

25. Kurt Zenz House, "The Limits of Energy Storage Technology," *Bulletin of the Atomic Scientists*, January 20, 2009, http://thebulletin.org/limits-energy-storage-technology.

26. Florian Steinke, Philipp Wolfrum, and Clemens Hoffmann, "Grid vs. Storage in a 100% Renewable Europe," *Renewable Energy* 50 (February 2013): 826–32, doi:10.1016/j.renene.2012.07.044.

27. T. Mai, D. Sandor, R. Wiser, and T. Schneider, *Renewable Electricity Futures Study: Executive Summary* (Golden, CO: National Renewable Energy Laboratory, 2012), http://www.nrel.gov/docs/fy13osti/52409-ES.pdf.

28. Electric Power Research Institute, *Estimating the Costs and Benefits of the Smart Grid*, March 29, 2011, http://my.epri.com/portal/server.pt?Abstract_id=000000000001022519.

29. Lannis Kannberg et al., *GridWiseTM: The Benefits of a Transformed Energy System*, Pacific Northwest National Laboratory, (Springfield VA: U.S. Department of Commerce, September 2003), http://arxiv.org/pdf/nlin/0409035v1.pdf.

30. William Atkinson, "Beyond Deployment Smart Meter Maintenance, Repair and Replacement," *Intelligent Utility*, January/February 2009, http://www.intelligentutility.com/magazine/article/107546/beyond-deployment-smart-meter-maintenance-repair-and-replacement. See also K. T. Weaver, "Congressional Testimony: 'Smart' Meters Have a Life of 5 to 7 Years," *Smart Grid Awareness*, October 29, 2015, http://smartgridawareness.org/2015/10/29/smart-meters-have-life-of-5-to-7-years/.

31. Marco Silva, Hugo Morais, and Zita Vale, "An Integrated Approach for Distributed Energy Resource Short-Term Scheduling in Smart Grids Considering Realistic Power System Simulation," *Energy Conversion and Management* 64 (2012): 273–88, accessed October 3, 2015, http://www.sciencedirect.com/science/article/pii/S0196890412002087.

32. Elizabeth Boyle, "V2G Generates Electricity—and Cash," *University of Delaware UDaily*, December 9, 2007, http://www.udel.edu/PR/UDaily/2008/nov/car112807 .html.

33. Pekka E. Kauppi et al., "Returning Forests Analyzed with the Forest Identity," *Proceedings of the National Academy of Sciences* 103, no. 46 (2006): 17574–79.

34. REN21, *Renewables 2014 Global Status Report* (Paris: Ren21 Secretariat, 2014), 31– 37, http://www.ren21.net/status-of-renewables/global-status-report/.

35. REN21, *Renewables 2014 Global Status Report*, 13.

36. REN21, *Renewables 2014 Global Status Report*, 13.

37. The International Energy Agency estimates that the world can double hydroelectric output by 2050, https://www.iea.org/topics/renewables/subtopics/hydropower/, accessed October 1, 2015.

38. REN21, *Renewables 2014 Global Status Report*, 39.

39. On induced seismicity, see Geoscience Australia, *Induced Seismicity and Geothermal Power Development in Australia*, (undated), http://www.ga.gov.au/corporate _data/66220/66220.pdf.

40. REN21, *Renewables 2014 Global Status Report*, 38

41. Benjamin Matek, *2015 Annual U.S. & Global Geothermal Power Production Report*, Geothermal Energy Association (2015), 15, http://geo-energy.org/reports.aspx.

42. Idaho National Laboratory, *The Future of Geothermal Energy: Impact of Enhanced Geothermal Systems (EGS) on the United States in the 21st Century* (U.S. Department of Energy, November 2006), https://mitei.mit.edu/system/files/geothermal-energy-full.pdf. Adam Goldstein and Ralph Braccio, *2013 Market Trends Report: Geothermal Technologies Office* (U.S. Department of Energy, January 2014), vi, http://www1.eere.energy.gov/geothermal/pdfs/market-report2013.pdf.

43. See for example T. Mai et al., *Renewable Electricity Futures Study Volume 1: Exploration of High-Penetration Renewable Electricity Futures* (Golden, CO: National Renewable Energy Laboratory, 2012), http://www.nrel.gov/docs/fy12osti/52409-1 .pdf.

44. Lauren Frayer, "Tiny Spanish Island Nears Its Goal: 100 Percent Renewable Energy," *National Public Radio*, September 28, 2014, http://www.npr.org/sections /parallels/2014/09/17/349223674/tiny-spanish-island-nears-its-goal-100-percent -renewable-energy.

45. See, for example, The Solutions Project, http://thesolutionsproject.org/.

46. "Will Renewables Replace Fossil Fuels?," recorded discussion with Mark Jacobson, David Blittersdorf, and Tom Murphy, *The Energy Xchange*, September 1, 2015, https://energyx.org/will-renewables-replace-fossil-fuels.

47. Massachusetts Institute of Technology Energy Initiative, *The Future of Solar Energy* (2015), xii–xx, http://mitei.mit.edu/futureofsolar.

48. Michael Dale and Sally Benson, "Energy Balance of the Global Photovoltaic (PV) Industry: Is the PV Industry a Net Electricity Producer?"

49. Ugo Bardi, *Extracted: How the Quest for Mineral Wealth Is Plundering the Planet* (White River Jct., VT: Chelsea Green, 2014), 131.

50. Amanda Adams and David Keith. "Are Global Wind Power Resource Estimates Overstated?" *Environmental Research Letters* 8, no. 1 (2013): 015021, http://iopscience.iop.org/article/10.1088/1748-9326/8/1/015021.

51. Kate Marvel, Ben Kravitz, and Ken Caldeira, "Geophysical Limits to Global Wind Power," *Nature Climate Change* 3 (2013), 118–21, http://www.nature.com/nclimate/journal/v3/n2/full/nclimate1683.html.

52. Alice Salt, "Wind Turbines Can Be Hazardous to Human Health," Cochlear Fluids Research Laboratory, Washington University, St. Louis, April 2, 2014, http://oto2.wustl.edu/cochlea/wind.html.

53. Dave Levitan, "Is Anything Stopping a Truly Massive Build-Out of Desert Solar Power?" *Scientific American*, July 1, 2013, http://www.scientificamerican.com/article/challenges-for-desert-solar-power/.

54. Red Eléctrica de España, "The Spanish Electricity System 2014" (REE: Madrid, 2015), 11, http://www.ree.es/sites/default/files/downloadable/the_spanish_electricity_system_2014_0.pdf.

55. Fraunhofer ISE, "Annual Electricity Generation in Germany," accessed October 1, 2015, https://www.energy-charts.de/energy.htm.

56. BP, "Data Workbook—Statistical Review 2015," http://www.bp.com/en/global/corporate/energy-economics/statistical-review-of-world-energy/downloads.html.

57. Red Eléctrica de España, "The Spanish Electricity System 2014."

58. Toby Couture, "Booms, Busts, and Retroactive Cuts: Spain's RE Odyssey," *E3 Analytics*, February 2011, http://www.e3analytics.eu/wp-content/uploads/2012/05/Analytical_Brief_Vol3_Issue1.pdf.

59. Andres Cala, "Renewable Energy in Spain Is Taking a Beating," *New York Times*, October 8, 2013, http://www.nytimes.com/2013/10/09/business/energy-environment/renewable-energy-in-spain-is-taking-a-beating.html?_r=0; Toby Couture, "The Lesson in Renewable Energy Development from Spain," *Renewable Energy World*, July 30, 2013, http://www.renewableenergyworld.com/rea/news/article/2013/07/a-lesson-in-renewable-energy-development-from-spain.

60. Harry Wirth, ed., *Recent Facts about Photovoltaics in Germany* (Freiburg: Fraunhofer ISE, 2015), 10, https://www.ise.fraunhofer.de/en/publications/veroeffentlichungen-pdf-dateien-en/studien-und-konzeptpapiere/recent-facts-about-photovoltaics-in-germany.pdf.

61. Bruno Burger, *Electricity Production from Solar and Wind in Germany in 2014* (Freiburg: Fraunhofer ISE, December 29, 2014), https://www.ise.fraunhofer.de/en/renewable-energy-data.

62. Kiley Kroh, "Germany Sets New Record, Generating 74% of Power Needs from Renewable Energy," *Climate Progress*, May 13, 2014, http://thinkprogress.org/climate/2014/05/13/3436923/germany-energy-records/Germany/.

63. Craig Morris, "Rebuttal: Renewables Make Millions of Germans Multidozenaires," *Renewables International*, May 16, 2014, http://www.renewablesinternational.net/renewables-make-millions-of-germans-multidozenaires/150/537/78900/.

64. California Energy Commission and California Public Utilities Commission, "California Solar Statistics: Program Totals by Administrator," accessed October 25, 2015, https://www.californiasolarstatistics.ca.gov/reports/agency_stats/.

65. Solar Server, "Energy Transition 2.0: Energy Storage and Solar PV," October 15, 2013, http://www.solarserver.com/solar-magazine/solar-report/solar-report/energy-transition-20-energy-storage-and-solar-pv.html.

66. Matthew Karnitschnig, "Germany's Expensive Gamble on Renewable Energy," *Wall Street Journal*, August 26, 2014, http://www.wsj.com/articles/germanys-expensive-gamble-on-renewable-energy-1409106602.

67. World Bank, World Development Indicators, "Industry, Value Added (% of GDP)," accessed October 1, 2015, http://data.worldbank.org/indicator/NV.IND.TOTL.ZS.

68. Christian Roselund and John Bernhardt, "Lessons Learned along Europe's Road to Renewables," *IEEE Spectrum*, May 4 2015, http://spectrum.ieee.org/energy/renewables/lessons-learned-along-europes-road-to-renewables.

69. Christopher Helman, "Will Solar Cause a Death Spiral for Utilities?," *Forbes*, January 30, 2015, http://www.forbes.com/sites/energysource/2015/01/30/will-solar-cause-a-death-spiral-for-utilities/.

70. Janine Schmidt, "Renewable Energies and Base Load Power Plants: Are They Compatible?," *Renews Special* 35 (June 2010), German Renewable Energies Agency, http://www.unendlich-viel-energie.de/media/file/302.35_Renews_Special_Renewable_Energies_and_Baseload_Power_Plants.pdf. See also Chris Nelder, "Why Base Load Power Is Doomed," *ZDnet*, March 28, 2012, http://www.zdnet.com/article/why-baseload-power-is-doomed/.

71. Joby Warrick, "Utilities Wage Campaign against Rooftop Solar," *Washington Post*, March 7, 2015, http://www.washingtonpost.com/national/health-science/utilities-sensing-threat-put-squeeze-on-booming-solar-roof-industry/2015/03/07/2d916f88-c1c9-11e4-ad5c-3b8ce89f1b89_story.html.

72. Joby Warrick, "Utilities Wage Campaign against Rooftop Solar."

Chapter 4

1. One barrel of oil = 5.7 million BTU, or 1670 kWh. The average human works at a power output of about 70 W. Multiplied by an 8–9 hour workday a person produces about 0.6 kWh of work per day; 1670 kWh ÷ 0.6 kWh per day = 2833 days. At 250 workdays per year, this equals about 11 years.

2. Lawrence Livermore National Laboratory, "Estimated U.S. Energy Use in 2014: ~ 98.3 Quads," accessed October 1, 2015, https://flowcharts.llnl.gov/content /assets/images/energy/us/Energy_US_2014.png.

3. See Rose George, *Ninety Percent of Everything* (New York: Metropolitan Books, 2013).

4. Danielle Murray, "Oil and Food: A Rising Security Challenge," *Earth Policy Institute,* May 9, 2005, http://www.earth-policy.org/index.php?/plan_b_updates/2005/ update48. Martin Heller and Gregory Keoleian. "Assessing the sustainability of the U.S. Food System: a Life Cycle Perspective," *Agricultural Systems* 76, no. 3 (2003): 1007–41. According to Murray, 21 percent of all food system energy goes to agriculture, and 34 percent of energy used in agriculture is gasoline and diesel; thus 7.1 percent of all food system energy is oil used in agriculture; add to that 14 percent of food system energy that is consumed in transport (presumably nearly all oil) for a total of 21 percent.

5. James Ayre, "Electric Car Demand Growing, Global Market Hits 740,000 Units," *Clean Technica,* March 28, 2015, http://cleantechnica.com/2015/03/28/ev-demand -growing-global-market-hits-740000-units/.

6. James Ayre, "Electric Car Demand Growing."

7. Sérgio Faias et al, "Energy Consumption and CO2 Emissions Evaluation for Electric and Internal Combustion Vehicles Using a LCA Approach," paper presented at the International Conference on Renewable Energies and Power Quality, La Coruna, Spain, March 25–27, 2015, http://www.icrepq.com/icrepq'11/660-faias .pdf.

8. BP, *Statistical Review of World Energy* (annual), http://bp.com/statisticalreview.

9. "Energy and Road Construction—What's the Mileage of Roadway?" *Pavement Interactive,* February 21, 2012, http://www.pavementinteractive.org/2012/02/21 /energy-and-road-construction-whats-the-mileage-of-roadway/.

10. Central Intelligence Agency, "The World Factbook, Field Listing: Roadways," accessed October 1, 2015, https://www.cia.gov/Library/publications/the-world -factbook/fields/2085.html.

11. U.S. Department of Energy Alternative Fuels Data Center, "Alternative Fuels Data Center—Fuel Properties Comparison," accessed October 1, 2015, http://www.afdc .energy.gov/fuels/fuel_comparison_chart.pdf). U.S. Department of Energy Alternative Fuels Data Center, "Charging Plug-In Electric Vehicles at Home," accessed October 1, 2015, http://www.afdc.energy.gov/fuels/electricity_charging_home .html. 33.70 kWh has 100% of the energy of one gallon of gasoline. Assuming a typical sedan in 2014 fuel economy of 28 miles per gallon, 28 ÷ 33.70 = 0.83 miles per kWh for a gasoline-powered sedan. An electric sedan in 2014 got 100 miles per 34 kWh, or 2.94 miles per kWh.

12. U.S. Department Transportation, Bureau of Transportation Statistics, *Freight Facts and Figures 2013* (January 2014), http://www.ops.fhwa.dot.gov/freight/freight_analysis/nat_freight_stats/docs/13factsfigures/pdfs/fff2013_highres.pdf.

13. Sean Kilgar, "Rolling Down That Electric Highway," *Fleet Owner*, August 13, 2014, accessed October 1, 2015, http://fleetowner.com/blog/rolling-down-electric-highway.

14. See, for example, Josie Garthwaite, "Car2go, Daimler-Backed Sharing Program, to Go Electric in San Diego," *New York Times* Wheels blog, July 13, 2011, http://wheels.blogs.nytimes.com/2011/07/13/car2go-daimler-backed-sharing-program-to-go-electric-in-san-diego/.

15. U.S. Energy Information Administration, "Frequently Asked Questions: How Much Ethanol Is Produced, Imported, and Consumed in the United States?" accessed October 1, 2015, http://www.eia.gov/tools/faqs/faq.cfm?id=90&t=4. U.S. Energy Information Administration, "Table 1. U.S. Biodiesel Production Capacity and Production," accessed October 1, 2015, http://www.eia.gov/biofuels/biodiesel/production/table1.pdf.

16. U.S. Energy Information Administration, "Frequently Asked Questions: How Much Gasoline Does the United States Consume?" accessed October 1, 2015, https://www.eia.gov/tools/faqs/faq.cfm?id=23&t=10.

17. Air and Transport Action Group, "Facts and Figures," accessed October 1, 2015, http://www.atag.org/facts-and-figures.html.

18. LaznaTech, "Technical Overview," accessed October 1, 2015, http://www.lanzatech.com/innovation/technical-overview/.

19. Sean Buchanan, "European Biofuel Bubble Bursts," *Inter Press Service News Agency*, April 28, 2015, http://www.ipsnews.net/2015/04/european-biofuel-bubble-bursts/.

20. U.S. Energy Information Administration, "Cellulosic Biofuels Begin to Flow but in Lower Volumes than Foreseen by Statutory Targets," *Today in Energy*, February 26, 2013, http://www.eia.gov/todayinenergy/detail.cfm?id=10131.

21. Jim Lane, "Where Are We with Algae Biofuels?" *Biofuels Digest*, October 13, 2014, http://www.biofuelsdigest.com/bdigest/2014/10/13/where-are-we-with-algae-biofuels/.

22. U.S. Energy Information Administration, "International Energy Statistics: Consumption of Jet Fuel, 2013, accessed October, 1, 2015, http://www.eia.gov/cfapps/ipdbproject/IEDIndex3.cfm.

23. U.S. Department of Agriculture, "Global Production of Vegetable Oils from 2000/01 to 2014/15 (in million metric tons)," via Statista, accessed October 1, 2015, http://www.statista.com/statistics/263978/global-vegetable-oil-production-since-2000-2001/.

24. Charles Hall, Stephen Balogh, and David Murphy, "What Is the Minimum EROI That a Sustainable Society Must Have?" *Energies* 2, no. 1 (2009): 25–47. Charles

Hall, Bruce Dale, and David Pimentel, "Seeking to Understand the Reasons for Different Energy Return on Investment (EROI) Estimates for Biofuels," *Sustainability* 3, no. 12 (2011): 2413–32, http://www.mdpi.com/2071-1050/3/12/2413/htm.

25. Jason M. Townsend et al., "Energy Return on Investment (EROI), Liquid Fuel Production, and Consequences for Wildlife," in *Peak Oil, Economic Growth, and Wildlife Conservation*, ed. Brian Czech and J. Edward Gates (New York: Springer, 2014), 29–61, accessed, October 2015, http://link.springer.com/chapter/10.1007/978-1-4939-1954-3_2.

26. C. Matthew Rendleman and Hosein Shapouri, "New Technologies in Ethanol Production," U.S. Department of Agriculture, Agricultural Economic Report Number 842 (February 2007), 5, http://www.usda.gov/oce/reports/energy/aer842_ethanol.pdf.

27. Nuria Basset et al., "The Net Energy of Biofuels," Erasmus Intensive Program: Energy Production from Biomass in the European Union (June 2010), http://www.iperasmuseprobio.unifg.it/dwn/THENETENERGYOFBIOFUELS.pdf.

28. Johanna Ivy, *Summary of Electrolytic Hydrogen Production: Milestone Completion Report* (Golden, CO: National Renewable Energy Lab, September 2004), 8, http://www.nrel.gov/docs/fy04osti/36734.pdf.

29. See, for example, National Renewable Energy Laboratory, "Hydrogen and Fuel Cell Research," accessed October 1, 2015, http://www.nrel.gov/hydrogen/proj_production_delivery.html.

30. Matthew Pellow et al., "Hydrogen or Batteries for Grid Storage? A Net Energy Analysis," *Energy and Environmental Science* 8 (2015), 1938–52, doi:10.1039/C4EE04041D.

31. "The Dawn of Hydrogen," *Ship and Bunker*, August 6, 2013, http://shipandbunker.com/news/features/fathom-spotlight/297814-the-dawn-of-hydrogen.

32. J. David Hughes, *Drilling Deeper: A Reality Check on U.S. Government Forecasts for a Lasting Tight Oil and Shale Boom* (Santa Rosa, CA: Post Carbon Institute, 2014), http://www.postcarbon.org/publications/drillingdeeper/. J. David Hughes, *Shale Gas Reality Check: Revisiting the U.S. Department of Energy Play-by-Play Forecasts through 2040 from Annual Energy Outlook 2015* (Santa Rosa, CA: Post Carbon Institute, 2015), http://www.postcarbon.org/publications/shale-gas-reality-check/.

33. Werner Zittel et al., "Fossil and Nuclear Fuels—the Supply Outlook," (Berlin: Energy Watch Group, March 2013), http://energywatchgroup.org/wp-content/uploads/2014/02/EWG-update2013_short_18_03_2013.pdf.

34. Hellenic Shipping News Worldwide, "Bunker Fuel Industry: China and Singapore Account Majority of Global Bunker Fuel Consumption," August 6, 2015, http://www.hellenicshippingnews.com/bunker-fuel-industry-china-and-singapore-account-majority-of-global-bunker-fuel-consumption/.

35. Katharine Sanderson, "Ship Kites in to Port," *Nature* (February 8, 2008), doi:10.1038/news.2008.564. Jan Lundberg, "Marine Sail Freight—America Gets Serious about Clean, Renewable Energy for Transport," *Sail Transport Network*, September 15, 2015, http://www.sailtransportnetwork.org/node/955.
36. See Sail Transport Network, http://sailtransportnetwork.org/.
37. Julian Jackson, "Sail Transport Network—the Past Meets the Future," *Earth Times*, June 10, 2011, http://www.earthtimes.org/scitech/sail-transport-network-the-past-meets-future/1010/#sthash.ZGHtV2w0.dpuf.
38. Julian Jackson, "Sail Transport Network."
39. International Energy Agency, *Special Report: World Energy Investment Outlook* (Paris, 2014), https://www.iea.org/publications/freepublications/publication/WEIO2014.pdf.

Chapter 5

1. J. Mukerji, "Refractories in Cement Making," in *Advances in Cement Technology: Critical Reviews and Case Studies on Manufacturing, Quality Control,Optimization and Use*, ed. S. N. Ghosh (New York: Pergamon Press, 1983), 265–86.
2. U.S. Energy Information Administration, "The Cement Industry Is the Most Energy Intensive of All Manufacturing Industries," *Today in Energy*, July 1, 2013, http://www.eia.gov/todayinenergy/detail.cfm?id=11911.
3. Kris De Decker, "The Bright Future of Solar Thermal Powered Factories," *Low Tech Magazine*, July 2011, http://www.lowtechmagazine.com/2011/07/solar-powered-factories.html.
4. International Energy Agency, *Energy Efficiency Indicators for Public Electricity Production from Fossil Fuels* (Paris, 2008), http://www.iea.org/publications/freepublications/publication/En_Efficiency_Indicators.pdf.
5. Kris De Decker, "The Bright Future of Solar Thermal Powered Factories."
6. Kris De Decker, "The Bright Future of Solar Thermal Powered Factories."
7. Clara Smith, Rich Belles, and A. J. Simon, *2007 Estimated International Energy Flows* (Lawrence Livermore National Laboratory, 2011), https://flowcharts.llnl.gov/content/international/2007EnergyInternational.pdf.
8. World Steel Association, "Crude Steel Production 2014–2015," accessed September 27, 2015, https://www.worldsteel.org/statistics/crude-steel-production.html.
9. Laura Sonter et al., "Carbon Emissions Due to Deforestation for the Production of Charcoal Used in Brazil's Steel Industry," *Nature Climate Change* 5 (2015): 359–63; doi:10.1038/nclimate2515.
10. Anna Simet, "World Bioenergy Association Releases Biogas Fact Sheet," *Biomass Magazine*, June 5, 2013, http://biomassmagazine.com/articles/9061/world-bioenergy-association-releases-biogas-fact-sheet.

11. Marc Jacobson et al., "100% Clean and Renewable Wind, Water, and Sunlight (WWS) All-Sector Energy Roadmaps for the 50 United States," *Energy & Environmental Science* 8, no. 7 (July 2015): 2094, doi:10.1039/c5ee01283j.

12. Volker Hoenig, Helmut Hoppe, and Bernhard Emberger, *Carbon Capture Technology—Options and Potentials for the Cement Industry* (Düsseldorf, Germany: European Cement Research Academy, 2007), http://www.ecra-online.org/fileadmin /redaktion/files/pdf/ECRA_Technical__Report_CCS_Phase_I.pdf.

13. REN21, *Renewables 2014 Global Status Report* (Paris: REN21 Secretariat, 2014), 15.

14. REN21, *Renewables 2014 Global Status Report*, 54.

15. Claudia Vannoni, Riccardo Battisti, and Serena Drigo, *Potential for Solar Heat in Industrial Processes* (Madrid: CIEMAT, 2008), http://www.aee-intec.at/0uploads /dateien561.pdf.

16. Werner Weiss and Matthias Rommel, *Medium Temperature Collectors* (Gleisdorf: AEE INTEC, 2005), http://www.aee-intec.at/0uploads/dateien86.pdf.

17. GeoDH, "What Is Geothermal District Heating?," accessed September 28, 2015, http://geodh.eu/about-geothermal-district-heating/.

18. REN21, *Renewables 2014 Global Status Report*, 28.

19. See Passive House Institute U.S., http://www.phius.org.

20. International Energy Agency, *Key World Energy Statistics* (Paris: OECD/IEA, 2014).

21. European Bioplastics, "Production Capacity," accessed September 29, 2015, http: //en.european-bioplastics.org/market/market-development/production-capacity/. Worldwatch Institute, "Global Plastic Production Rises, Recycling Lags," January 28, 2015, http://www.worldwatch.org/global-plastic-production-rises-recycling -lags-0.

22. Dan Charles, "Fertilized World," *National Geographic*, May 2013, http://ngm .nationalgeographic.com/2013/05/fertilized-world/charles-text.

23. Organic Consumers Association, "Organic and Sustainable Farmers Can Feed the World," accessed September 29, 2015, https://www.organicconsumers.org/news /organic-and-sustainable-farmers-can-feed-world.

24. "Energy and Road Construction—What's the Mileage of Roadway?" *Pavement Interactive*, February 21, 2012, http://www.pavementinteractive.org/2012/02/21 /energy-and-road-construction-whats-the-mileage-of-roadway/.

25. Designboom, "'sand.stone.road' by thomas kosbau + andrew wetzler iida awards 2010 winner," accessed September 30, 2015, http://www.designboom.com/design /sandstoneroad-by-thomas-kosbau-andrew-wetzler-iida-awards-2010-winner/.

26. Kieron Monks, "Would You Live in a House Made of Sand and Bacteria? It's a Surprisingly Good Idea," *CNN*, May 22, 2014, http://www.cnn.com/2014/05/21/tech /innovation/would-you-live-in-a-house-made-of-urine-and-bacteria/.

27. Morgana Matus, "NASA Harnessing the Power of Microbes to Create Building Bricks on Mars," *Inhabitat*, October 5, 2012, http://inhabitat.com/nasa-harnessing -the-power-of-microbes-to-create-bricks-and-mortar-on-mars/.

28. Sunny Soni and Madhu Agarwal, "Lubricants from Renewable Energy Sources—a Review," *Green Chemistry Letters and Reviews* 7, no. 4 (2014): 359–82, doi:10.1080/17518253.2014.959565.

Chapter 6

1. International Energy Agency (IEA), "Scenarios and Projections," accessed September 30, 2015, http://www.iea.org/publications/scenariosandprojections/.
2. International Energy Agency, *Key World Energy Statistics*, (Paris: OECD/IEA, 2014).
3. Jessica Lambert et al., "Energy, EROI and Quality of Life," *Energy Policy* 64 (2014): 153–67, doi:10.1016/j.enpol.2013.07.001.
4. Ida Kubiszewski, Cutler Cleveland, and Peter Endres, "Meta-analysis of Net Energy Return for Wind Power Systems," *Renewable Energy* 35 (2010): 218–25, doi:10.1016/j.renene.2009.01.012.
5. Marco Raugei, Pere Fullana-i-Palmer, and Vasilis Fthenakis, "The Energy Return on Energy Investment (EROI) of Photovoltaics: Methodology and Comparisons with Fossil Fuel Life Cycles," *Energy Policy* 45 (June 2012): 576–82, doi:10.1016/j.enpol.2012.03.008; https://www.bnl.gov/pv/files/pdf/241_Raugei_EROI_EP_revised_II_2012-03_VMF.pdf.
6. Pedro Prieto and Charles A. S. Hall, *Spain's Photovoltaic Revolution: The Energy Return on Investment* (New York: Springer, 2011).
7. Graham Palmer, *Energy in Australia: Peak Oil, Solar Power, and Asia's Economic Growth* (New York: Springer, 2014).
8. Michael Carbajales-Dale et al., "Energy Return on Investment (EROI) of Solar PV": An Attempt at Reconciliation," *Proceedings of the IEEE* 103, no. 7 (2015): 995–99, doi:10.1109/JPROC.2015.2438471.
9. Michael Carbajales-Dale et al., "Energy Return on Investment (EROI) of Solar PV.
10. Khagendra P. Bhandari et al., "Energy Payback Time (EPBT) and Energy Return on Energy Invested (EROI) of Solar Photovoltaic Systems: A Systematic Review and Meta-analysis," *Renewable and Sustainable Energy Reviews* 47 (2015): 133–41, doi:10.1016/j.rser.2015.02.057.
11. Ida Kubiszewski, Cutler Cleveland, and Peter Endres, "Meta-analysis of Net Energy Return for Wind Power Systems."
12. D. Weissbach et al., "Energy Intensities, EROIs (Energy Returned on Invested), and Energy Payback Times of Electricity Generating Power Plants," *Energy* 52 (2013): 210–21, accessed October 2, 2015, doi:10.1016/j.energy.2013.01.029.
13. John Weber, "Machines Making Machines Making Machines," *sunweber* blog, December 3, 2011, http://sunweber.blogspot.com/2011/12/machines-making-machines-making.html.

14. The Sahara Solar Breeder Foundation has plans along these lines, but it is unclear what stage they have achieved. http://www.ssb-foundation.com/, accessed October 2, 2015.

15. Kris De Decker, "How Sustainable is PV Solar Power?," *Low Tech Magazine*, April 2015, http://www.lowtechmagazine.com/2015/04/how-sustainable-is-pv-solar-power .html.

16. Mark Z. Jacobson et al., "100% Clean and Renewable Wind, Water, and Sunlight (WWS) all-sector energy roadmaps for the 50 United States," *Energy and Environmental Science* 8, no. 7 (2015): 2093–2117.

17. Daniel J. Graeber, "Support for Renewables Lacking, Global Reports Find," *UPI*, May 19, 2015, http://www.upi.com/Business_News/Energy-Industry/2015/05/19 /Support-for-renewables-lacking-global-reports-find/1711432033175/.

18. Mark Z. Jacobson and Mark A. Delucchi, "A Plan to Power 100 Percent of the Planet with Renewables," *Scientific American*, (November 1, 2009), http://www .scientificamerican.com/article/a-path-to-sustainable-energy-by-2030/.

19. Frankfurt School–UNEP Collaborating Centre for Climate & Sustainable Energy Finance (FS-UNEP) and Bloomberg New Energy Finance, *Global Trends in Renewable Energy Investment 2015–Chart Pack* (Frankfurt: FS-UNEP, 2015), http: //fs-unep-centre.org/sites/default/files/attachments/unep_fs_globaltrends2015_ chartpack.pdf. International Energy Agency (IEA), *World Energy Investment Outlook 2014 Fact Sheet* (Paris: IEA, 2014).

20. Stockholm International Peace Research Institute (SIPRI), "Military Expenditure," accessed October 2, 2015, http://www.sipri.org/research/armaments/milex.

21. Richard Heinberg, *The End of Growth: Adapting to Our New Economic Reality* (Gabriola Island, BC: New Society Publishers, 2011).

22. World Bank, "World Bank Open Data," accessed September 7, 2015, http://data. worldbank.org/.

23. See the work of Emmanuel Saez at University of California–Berkeley, http://eml .berkeley.edu/~saez/#income.

24. Alexander Gloy, "Analyze This—the Fed Is Not Printing Enough Money!," *Zero Hedge*, September 8, 2012, http://www.zerohedge.com/news/guest-post-analyze -fed-not-printing-enough-money.

25. Based on median household income in the United States and the United Kingdom: Office for National Statistics (ONS), *Middle Income Households, 1977–2010/11* (London: ONS, 2013). Carmen DeNavas-Walt and Bernadette D. Proctor, *Income and Poverty in the United States: 2014* (Washington: U.S. Census Bureau, 2015).

26. International Energy Agency, *Energy Efficiency Indicators for Public Electricity Production from Fossil Fuels* (Paris, 2008), http://www.iea.org/publications /freepublications/publication/En_Efficiency_Indicators.pdf.

27. International Energy Agency, *Key World Energy Statistics* (Paris: OECD/IEA, 2014).

28. "Improving IC Engine Efficiency," *University of Washington*, accessed October 2, 2015, http://courses.washington.edu/me341/oct22v2.htm.

29. "Electrical Motor Efficiency," *Engineering Toolbox*, accessed October 2, 2015, http://www.engineeringtoolbox.com/electrical-motor-efficiency-d_655.html.

30. Ryan Carlyle, "If All U.S. Cars Suddenly Became Electric, How Much More Electricity Would We Need?" *Slate*, May 2, 2014, http://www.slate.com/blogs/quora/2014/05/02/electric_vehicles_how_much_energy_would_we_need_to_fuel_them.html.

31. This figure is difficult to calculate globally. Space conditioning accounts for about 45 percent of residential energy use in the United States, less in Europe, and even less in China and India. The U.S. Energy Information Administration shows about 52 quadrillion BTU of site use energy in the residential sector, or 92 counting electricity generation losses (http://www.eia.gov/tools/faqs/faq.cfm?id=447&t=1). Assuming that space conditioning is about 30 percent site use on average, then the reduction would be about 25 quadrillion BTU or 620 Mtoe. But not all of this is in the form of electricity, so starting at 92 overstates the savings. On the basis of site energy, the savings would be about 350 Mtoe.

32. Ilhan Ozturk, et all., "Energy consumption and economic growth relationship: Evidence from panel data for low and middle income countries," *Energy Policy 38*, no. 8 (2010): 4422–28, doi:10.1016/j.enpol.2010.03.071.

33. A significant problem with economic intensity measurements as done today, is that they include GDP from all sources (including trade), but energy use from only within the national boundaries. That is, the embodied energy in imports is not counted toward energy consumption (just as embodied energy of exports is not deducted from energy used). For a large importer such as the United States, the offshoring of manufacturing has led to efficiency improvements economy-wide that may be totally offset by the lower efficiency of production in China, for example. See Thomas Wiedmann et al., "The Material Footprint of Nations," *Proceedings of the National Academy of Sciences of the United States of America* 112, no. 20 (May 19, 2015), 6271-6276, doi:10.1073/pnas.1220362110.

34. Enerdata, "Energy Intensity of GDP at Constant Purchasing Power Parities," *Global Energy Statistical Yearbook 2015*, https://yearbook.enerdata.net/energy-intensity-GDP-by-region.html.

35. Thomas Wiedmann et al., "The Material Footprint of Nations."

36. Jesse Jenkins and Armond Cohen, "The Role of Energy Intensity in Global Decarbonization: How Fast Can We Cut Energy Use?" *The Energy Collective*, March 16, 2015, http://www.theenergycollective.com/jessejenkins/2205386/role-energy-intensity-global-decarbonization-how-fast-can-we-cut-energy-use.

37. Peter Loftus et al., "A Critical Review of Global Decarbonization Scenarios: What Do They Tell Us about Feasibility?" *Wiley Interdisciplinary Reviews: Climate Change* 6, no. 1 (2015): 93–112.

38. See also Jenkins and Cohen, "The Role of Energy Intensity in Global Decarbonization," and Schalk Cloete, "Can We Really Decouple Living Standards from Energy Consumption?" *The Energy Collective*, June 8, 2015, http://www.theenergy collective.com/schalk-cloete/2231916/can-we-really-uncouple-welfare-growth -energy-growth.

39. PriceWaterhouse Coopers, *Two Degrees of Separation: Ambition and Reality; Low Carbon Economy Index 2014* (September 2014), http://www.pwc.co.uk/assets/pdf /low-carbon-economy-index-2014.pdf.

40. Kevin Anderson, "Avoiding Dangerous Climate Change Demands De-growth Strategies from Wealthier Nations," *kevinanderson.info*, November 25, 2013, http://kevinanderson.info/blog/avoiding-dangerous-climate-change-demands -de-growth-strategies-from-wealthier-nations/.

41. United Nations, Framework Convention on Climate Change, *Adoption of the Paris Agreement*, FCCC/CP/2015/L.9 (December 12, 2015), http://unfccc.int/resource /docs/2015/cop21/eng/l09.pdf.

42. Intergovernmental Panel on Climate Change (IPCC), *Climate Change 2014: Mitigation of Climate Change. Contribution of Working Group III to the Fifth Assessment Report of the Intergovernmental Panel on Climate Change* (Cambridge: Cambridge University Press, 2014); https://www.ipcc.ch/report/ar5/wg3/.

43. Intergovernmental Panel on Climate Change (IPCC), *Climate Change 2014: Mitigation of Climate Change. Contribution of Working Group III*.

44. Donella Meadows et al., *The Limits to Growth* (New York: Potomac Associates, 1972). Donella Meadows, Jorgen Randers, and Dennis Meadows, *Limits to Growth: The 30-Year Update* (White River Junction, VT: Chelsea Green, 2004).

45. Peter Victor, *Managing Without Growth: Slower by Design, Not Disaster* (Cheltenham, UK: Edward Elgar, 2008).

Chapter 7

1. Mark Z. Jacobson et al., "100% Clean and Renewable Wind, Water, and Sunlight (WWS) all-sector energy roadmaps for the 50 United States," *Energy and Environmental Science* 8, no. 7 (2015): 2093–2117."

2. Michael Brune, "All In," *Sierra Magazine*, January/February 2014, http://www .sierraclub.org/sierra/2014-1-january-february/feature/all.

3. Werner Zittel, Jan Zerhusen, and Martin Zerta, *Fossil and Nuclear Fuels—the Supply Outlook* (Ottobrunn, Germany: Energy Watch Group, 2013), http://energywatchgroup.org/wp-content/uploads/2014/02/EWG-update2013_long_18_03_2013up1.pdf.

4. Gavin Mudd, "The Future of Yellowcake: A Global Assessment of Uranium Resources and Mining," *Science of the Total Environment* 472 (2014): 590–607, doi:10.1016/j.scitotenv.2013.11.070.

5. Charles Hall and Bobby Powers, "The Energy Return of Nuclear Power. (EROI on the Web—Part 4 of 5)," *Oil Drum*, April 22, 2008, http://www.theoildrum.com/node/3877.

6. Dino Grandoni, "Why It's Taking the U.S. So Long to Make Fusion Energy Work," *Huffington Post*, January 20, 2015, http://www.huffingtonpost.com/2015/01/20/fusion-energy-reactor_n_6438772.html.

7. International Energy Agency, *Cost and Performance of Carbon Dioxide Capture from Power Generation* (Paris: OECD/IEA, 2011), http://www.iea.org/publications/freepublications/publication/cost-and-performance-of-carbon-dioxide-capture-from-power-generation.html.

8. This is our own calculation, based on the following sources. Density of compressed CO_2 in CCS: David McCollum and Joan Ogden, *Techno-economic Models for Carbon Dioxide Compression, Transport, and Storage & Correlations for Estimating Carbon Dioxide Density and Viscosity*, Institute for Transportation Studies, University of California–Davis, October 2006. Consumption of coal: U.S. Energy Information Administration, "Table 32. U.S. Coal Consumption by End-Use Sector, 2008–2014," accessed October 3, 2015, http://www.eia.gov/coal/production/quarterly/pdf/t32p01p1.pdf. Coal emissions factor: U.S. Energy Information Administration, "Carbon Dioxide Emissions Coefficients," accessed October 3, 2015, http://www.eia.gov/environment/emissions/co2_vol_mass.cfm.

9. This source (the U.S. Energy Information Administration's *Annual Energy Outlook 2015*) projects a 52 percent average premium of "advanced coal" with CCS over conventional coal, in terms of the levelized cost of electricity in $/MWh, http://www.eia.gov/forecasts/aeo/pdf/electricity_generation.pdf (see table 1). Table 2 shows minimum and maximum ranges, from 12 to 84 percent for coal CCS versus coal.

10. Richard Heinberg and Howard Herzog, "Does 'Clean Coal' Technology Have a Future?" *Wall Street Journal*, November 23, 2014, http://www.wsj.com/articles/does-clean-coal-technology-have-a-future-1416779351.

11. Jack Kittredge, *Soil Carbon Restoration: Can Biology Do the Job?* (Northeast Organic Farming Association, August 2015), http://www.nofamass.org/content/soil-carbon-restoration-can-biology-do-job.

12. R. A. Houghton, Brett Byers, and Alexander Nassikas, "A Role for Tropical Forests in Stabilizing Atmospheric CO_2," *Nature Climate Change* 5 (2015): 1022–23, http://www.nature.com/nclimate/journal/v5/n12/full/nclimate2869.html.

13. Jamie Lendino, "This Fully Transparent Solar Cell Could Make Every Window and Screen a Power Source (updated)," *ExtremeTech*, April 20, 2015, http://www.extremetech.com/extreme/188667-a-fully-transparent-solar-cell-that-could-make-every-window-and-screen-a-power-source.

14. Unlike microprocessors, which have a short design cycle and short lifetime on average, macrolevel technology is completely different. Coauthor David Fridley conducted an informal survey of his colleagues at Lawrence Berkeley National Laboratory on their estimation of the average time from benchtop proof-of-concept to commercialization with market impact; the answers averaged 25 to 30 years. This is true for many technologies: the first U.S. patent for solar cells was issued in 1888, Bell Labs was producing solar cells for the space program in the 1950s, and the first polysilicon solar cells as used today were produced in 1982. In other words, what we know now is pretty much what's available for us to build on; or, if it hasn't been invented yet, you're working on faith, not science. (This equally applies to the "black swan" type discoveries.)

15. Avaneesh Pandey, "Artificial Photosynthesis System Created From a Nanowire-Bacteria Hybrid," *International Business Times*, April 17, 2015, http://www.ibtimes.com/artificial-photosynthesis-system-created-nanowire-bacteria-hybrid-1886068.

16. Vaclav Smil, *Energy Transitions: History, Requirements, Prospects* (Santa Barbara: Praeger, 2010).

17. See, for example, Linda Federico-O'Murchu, "How 3-D Printing Will Radically Change the World," *CNBC*, May 11, 2014, http://www.cnbc.com/2014/05/09/will-3-d-technology-radically-change-the-world.html.

Chapter 8

1. United Nations, Department of Economic and Social Affairs, *The World Mortality Report 2013* (New York: United Nations, 2013). World Health Organization, *World Health Statistics 2015* (Geneva: WHO Press, 2015). IFAD, FAO, and WFP, "The State of Food Insecurity in the World 2015: Meeting the 2015 International Hunger Targets: Taking Stock of Uneven Progress." (Rome: FAO, 2015), http://www.fao.org/hunger/en/. Simon Fraser University School for International Studies, Human Security Research Group, *Human Security Report 2013: The Decline in Global Violence: Evidence, Explanation, and Contestation* (Vancouver: Human Security Research Group, 2013), http://hsrgroup.org/human-security-reports/2013/overview.aspx.

2. World Health Organization, "Household Air Pollution and Health," Fact Sheet No. 292, March 2014, http://www.who.int/mediacentre/factsheets/fs292/en/.

3. Brian Bienkowski, "Poor Communities Bear Greatest Burden from Fracking," *Scientific American*, May 6, 2015, http://www.scientificamerican.com/article/poor -communities-bear-greatest-burden-from-fracking/.

4. Wang Ming-Xiao et al., "Analysis of National Coal-Mining Accident Data in China, 2001–2008," *Public Health Reports* 126, no. 2 (March–April 2011): 270–75, http://www.ncbi.nlm.nih.gov/pmc/articles/PMC3056041/.

5. See Michael Watts, "Sweet and Sour: The Curse of Oil in the Niger Delta," in *The ENERGY Reader: Overdevelopment and the Delusion of Endless Growth*, ed. Tom Butler, Daniel Lerch, and George Wuerthner (Healdsburg, CA: Watershed Media, 2012), 247–55.

6. Ivan Illich, *Energy and Equity* (London: Calder & Boyars, 1974).

7. World Bank, "GINI Index (World Bank Estimate)," World Development Indicators, accessed September 5, 2015, http://data.worldbank.org/indicator/SI.POV .GINI.

8. E. F. Schumacher, *Small Is Beautiful: Economics as if People Mattered* (New York: Harper & Row, 1973).

9. Open Source Ecology, "Machines: Global Village Construction Set," accessed September 5, 2015, http://opensourceecology.org/gvcs/.

10. See http://www.growbiointensive.org/.

11. Helena Norberg-Hodge, *Ancient Futures: Learning from Ladakh* (San Francisco: Sierra Club Books, 1992).

12. Xie Yu and Zhou Xiang, "Income Inequality in Today's China," *Proceedings of the National Academy of Sciences of the United States of America* 111, no. 19 (May 13, 2014): 6928–33, http://www.pnas.org/content/111/19/6928.short. See also James Galbraith, "Global Inequality and Global Macroeconomics," *Journal of Policy Modeling* 29, no. 4 (July–August 2007): 587-607, http://www.sciencedirect.com/science /article/pii/S0161893807000452.

13. Lancet, "Editorial: (Barely) Living in Smog: China and Air Pollution," *Lancet* 383 (March 8, 2014): 845, http://www.thelancet.com/pdfs/journals/lancet /PIIS0140-6736(14)60427-X.pdf.

14. Of the many nongovernmental organizations working on these issues in the developing world, see especially the Institute for Transportation and Development Policy, https://www.itdp.org/.

15. Josh Bivens and Lawrence Mishel, "Understanding the Historic Divergence between Productivity and a Typical Worker's Pay," *Economic Policy Institute*, September 2, 2015, http://www.epi.org/publication/understanding-the-historic -divergence-between-productivity-and-a-typical-workers-pay-why-it-matters-and -why-its-real/.

16. See the work of Michael Shuman, notably *Local Economy Solution: How Innovative, Self-Financing "Pollinator" Enterprises Can Grow Jobs and Prosperity* (White River Junction, VT: Chelsea Green, 2015) and *Going Local: Creating Self-Reliant Communities in a Global Age* (New York: Simon & Schuster, 1998).

17. Paul Baer et al., *The Right to Development in a Climate Constrained World: The Greenhouse Development Rights Framework* (Berlin: Heinrich Böll Foundation, 2008), February 16, 2009, http://gdrights.org/2009/02/16/second-edition-of-the -greenhsouse-development-rights/.

18. Peter Barnes, "Common Wealth Trusts: Structures of Transition," *Great Transition Initiative*, August 2015, http://www.greattransition.org/publication/common -wealth-trusts.

19. The Norwegian sovereign wealth fund was funded almost completely by oil and gas development in the North Sea, yet it is divesting its holdings from coal, palm oil, and other industries and companies deemed environmentally unsound, highlighting an interesting paradox.

20. Alaska Permanent Fund Corporation, "About the Fund," accessed September 6, 2015, http://www.apfc.org/home/Content/aboutFund/aboutPermFund .cfm.

21. Texas Education Agency, "Texas Permanent School Fund," accessed September 6, 2015, http://tea.texas.gov/Finance_and_Grants/Permanent_School_Fund/.

22. California Air Resources Board, "Cap-and-Trade Program," accessed September 6, 2015, http://www.arb.ca.gov/cc/capandtrade/capandtrade.htm.

23. Melanie Curry, "California Cap-and-Trade Is 'Officially a Success,'" *Streetsblog California*, November 10, 2015, http://cal.streetsblog.org/2015/11/10/california -cap-and-trade-is-officially-a-success/.

24. See, for example, David Baker, "Pope Blasts California's Cap-and-Trade System," *SFGate*, June 18, 2015, http://www.sfgate.com/business/article/Pope-blasts-Cali- fornia-s-cap-and-trade-system-6336494.php, and Kate Sheppard, "Environmental Justice v. Cap-and-Trade," *American Prospect*, February 28, 2008, http://prospect .org/article/environmental-justice-v-cap-and-trade.

Chapter 9

1. Bill Gates, "We Need Energy Miracles," Gatesnotes, the Blog of Bill Gates, June 25, 2014, http://www.gatesnotes.com/Energy/Energy-Miracles.

2. International Energy Agency, *Clean Energy Progress Report: IEA Input to the Clean Energy Ministerial* (Paris: International Energy Agency, 2011), accessed on September 6, 2015, http://www.iea.org/publications/freepublications/publication/CEM _Progress_Report.pdf.

3. John Banusiewicz, U.S. Department of Defense, "Hagel to Address 'Threat Multiplier' of Climate Change," October 13, 2015, http://www.defense.gov/News -Article-View/Article/603440.

4. "Wind and Solar Energy Are Increasingly Competitive. But a Lot Has to Change Before They Can Make a Real Impact," *Economist* August 1, 2015, http://www .economist.com/news/leaders/21660124-wind-and-solar-energy-are-increasingly -competitive-lot-has-change-they-can-make.

5. Additional policies are gaining traction to advance local and regional targets. See Communities section in Chapter 10.

6. International Energy Agency, *Clean Energy Progress Report: IEA Input to the Clean Energy Ministerial* (Paris: International Energy Agency, 2011), 42–43, http: //www.iea.org/publications/freepublications/publication/CEM_Progress_Report .pdf.

7. For more reading on FIT design issues, as well as common myths and facts about Germany's energy transition, see: Diane Moss, *Power and Profits in the Hands of the People: Lessons Learned from Germany's Rural Renewable Energy Renaissance*, (Washington, DC: Heinrich Böll Stiftung, 2012), p. VI–VIII, http://bit.ly/moss-powerand; Diane Moss and Angelina Galiteva, "Clearing Up the Facts About Solar in Germany," Renewables 100 Policy Institute, February 16, 2012, http://bit.ly/re-100clearingup; Angelina Galiteva and Diane Moss, *Germany–California Learning and Collaboration Tour: Toward an Integrated Renewable Energy System*, (Washington DC: Renewables 100 Policy Institute, July 2014), p. 9-15 and 19-21, http://bit .ly/re100gercaltour; and Deutsche Bank Climate Change Advisors, *Paying for Renewable Energy: TLC at the Right Price*, December 2009, http://bit.ly/dbpayingforre.

8. International Energy Agency, *Clean Energy Progress Report*, 42-43.

9. U.S. Energy Information Agency, "Most states have Renewable Portfolio Standards," *Today In Energy*, February 3, 2012, http://www.eia.gov/todayinenergy /detail.cfm?id=4850.

10. Center for Climate and Energy Solutions, "Renewable and Alternative Energy Portfolio Standards," Map, accessed on September 5, 2015, http://www.c2es.org /node/9340.

11. J. Heeter et al., *A Survey of State-Level Cost and Benefit Estimates of Renewable Portfolio Standards* (U.S. Department of Energy, National Renewable Energy Laboratory and Lawrence Berkeley National Laboratory, May 2014), http://www.nrel.gov /docs/fy14osti/61042.pdf.

12. Legislative Analyst's Office (California), *The 2014–15 Budget: Cap-and-Trade Auction Revenue Expenditure Plan*, February 24, 2014, http://www.lao.ca .gov/reports/2014/budget/cap-and-trade/auction-revenue-expenditure-022414 .aspx.

13. David Flemming, *Energy and the Common Purpose, Descending the Energy Staircase with Tradable Energy Quotas (TEQs)*, (London: The Lean Economy Connection, 2006), http://www.teqs.net/EnergyAndTheCommonPurpose.pdf.

14. See www.renewables100.org and www.go100percent.org.

15. Yves Smith, "Military Spending Could Give Big Boost to Renewable Energy," *Naked Capitalism*, July 7, 2015, http://www.nakedcapitalism.com/2015/07/military -spending-could-give-big-boost -to-renewable-energy.html.

16. Citizen advocacy is often the impetus for these kinds of government actions. For example, the Bus Riders Union and Labor Community Strategy Center in Los Angeles are campaigning to flip national transport funding dollars from the current 80 percent roads/20 percent transit ratio to 20 percent roads/80 percent transit. http: //www.thestrategycenter.org/project/bus-riders-union.

17. See R. D. Van Buskirk et al., "A Retrospective Investigation of Energy Efficiency Standards: Policies May Have Accelerated Long Term Declines in Appliance Costs," *Environmental Research Letters* 9, no. 11 (November 13, 2014), http: //iopscience.iop.org/article/10.1088/1748-9326/9/11/114010/meta.

18. There are many examples and resources now for sustainability practices in city planning. See especially Congress for the New Urbanism (https://www.cnu.org/) and the "Sustaining Places" initiative of the American Planning Association, https://www.planning.org/sustainingplaces/compplanstandards/.

19. City of Portland and Multnomah County, *Climate Action Plan 2009: Year Two Progress Report* (April 2012), http://www.portlandoregon.gov/bps/article/393345.

20. Divya Pandey, Madhoolika Agrawal, and Jai Shanker Pandey, "Carbon Footprint: Current Methods of Estimation," *Environmental Monitoring and Assessment* 178, no. 1–4 (2011): 135–60, doi:10.1007/s10661-010-1678-y. Pandey et al. break carbon footprint analysis into three tiers: tier I is onsite fuel consumption, tier II considers emissions embodied in purchases of energy, tier III considers all other offsite emissions, such as international transport and production emissions in other countries. According to Pandey et al. tier III is not required by most GHG consultancies and is often omitted from carbon assessments (p. 145).

21. Thomas Wiedmann, "A Review of Recent Multi-region Input–Output Models Used for Consumption-Based Emission and Resource Accounting." *Ecological Economics* 69, no. 2 (2009): 211–22, doi:10.1016/j.ecolecon.2009.08.026. Wiedmann also used the term *Input–Output* or *I-O accounting*.

22. Pandey, "Carbon Footprint," 145. Consumption-based accounting falls under tier III.

23. World Maritime News, "COP21: Paris Remains Silent on Shipping and Aviation," December 14, 2015, http://worldmaritimenews.com/archives/178732/cop21-paris -remains-silent-on-shipping-and-aviation/.

24. David McCullum, Gregory Gould, and David Greene, *Greenhouse Gas Emissions from Aviation and Marine Transportation: Mitigation Potential and Policies* (Prepared for the Pew Center on Global Climate Change, Center for Climate and Energy Solutions, 2009), accessed on September 5, 2015, http://www.c2es.org/docUploads/aviation-and-marine-report-2009.pdf/.

25. While aviation emissions are not addressed in the Paris climate agreement, Europe has nevertheless begun regulating them (see http://ec.europa.eu/clima/policies/transport/aviation/index_en.htm). Indeed, it may be easier legally to regulate aviation emissions than to regulate embedded emissions in traded goods, the latter of which are regulated by the World Trade Organization.

Chapter 10

1. One such tool can be found at http://coolclimate.berkeley.edu/calculator.

2. Kris De Decker, "If We Insulate Our Houses, Why Not Our Cooking Pots?" *Low-Tech Magazine*, July 1, 2014, http://www.lowtechmagazine.com/2014/07/cooking-pot-insulation-key-to-sustainable-cooking.html.

3. Office of Energy Efficiency and Renewable Energy, "Top Ten Utility Green Power Programs (as of December 2014)," accessed on September 9, 2015, http://apps3.eere.energy.gov/greenpower/resources/tables/topten.shtml.

4. See Michal Shuman, *Local Dollars, Local Sense: How to Shift Your Money from Wall Street to Main Street and Achieve Real Prosperity* (White River Junction, VT: Chelsea Green, 2012).

5. Transition Network, accessed on September 9, 2015, http://www.transitionnetwork.org.

6. See http://climateprotection.org.

7. LEAN Energy U.S., "CCA by State," accessed on September 5, 2015, http://www.leanenergyus.org/cca-by-state/.

8. United States Department of Agriculture Office of Communications, "News Release: USDA Celebrates National Farmers Market Week, August 4–10," Release No. 0155.13, accessed on September 5, 2015, http://www.usda.gov//wps/portal/usda/usdamediafb?contentid=2013/08/0155.xml.

9. Sustainable Economics Law Center, "City Policies," accessed on September 9, 2015, http://www.theselc.org/city-policies.

Chapter 11

1. Kris De Decker, "How to Build a Low-Tech Internet," *Low-Tech Magazine*, October 25, 2015, http://www.lowtechmagazine.com/2015/10/how-to-build-a-low-tech-internet.html.

2. Kevin Anderson, "Duality in Climate Science," *Nature Geoscience* 8 (October 12, 2015): 898–900, http://www.nature.com/ngeo/journal/v8/n12/full/ngeo2559.html.

3. Global Footprint Network, *Living Planet Report 2014* (Gland, Switzerland: WWF, 2014), http://www.footprintnetwork.org/en/index.php/GFN/page/living_planet _report2/.

4. National Resources Defense Council, "NRDC Fact Sheet: California's Energy Efficiency Success Story," July 2013, http://www.nrdc.org/energy/files/ca-success -story-FS.pdf.

5. U.S. Bureau of Economic Analysis, "Table 1.1.5. Gross Domestic Product," accessed September 4, 2015, http://www.bea.gov/iTable/iTable.cfm?ReqID=9&step=1#reqid =9&step=3&isuri=1&903=5.

6. Richard Heinberg, "The Brief, Tragic Reign of Consumerism," *Green Social Thought* 64 (Spring 2014): 18–20; http://greensocialthought.org/wp-content /uploads/2014/07/gst64-18-20-Richard-Heinberg1.pdf.

7. See http://www.degrowth.org/.

8. See http://www.journals.elsevier.com/ecological-economics/.

9. United Nations, Department of Economic and Social Affairs, *The World Population Situation in 2014* (New York: United Nations, 2014).

10. See http://populationinstitute.org/.

11. See http://populationmedia.org/.

12. For a glimpse at what kind of energy has to be expended to transport and assemble a single wind turbine, it is worth watching this video of heavy vehicles and machinery doing the job in Ireland: http://bit.ly/transport-assemble-wind-turbine.

13. Michael Stone and Richard Heinberg, "You Can't Do Just One Thing: A Conversation with Richard Heinberg," Center for Ecoliteracy, May 9, 2012, http://www .ecoliteracy.org/article/you-cant-do-just-one-thing-conversation-richard-heinberg.

14. Susan Kraemer, "Zero Carbon Cement Production with Solar Thermal," *Clean Technica*, April 10, 2012, http://cleantechnica.com/2012/04/10/zero-carbon-cement -production-with-solar-thermal/. This article is an example, but the "cheaper to make" depends on the sale of the resultant carbon monoxide (CO), the price of which they cite is current. If the market were expanded with cement CO, then the price would drop dramatically since there would be over a billion metric tons of CO to sell.

About the Authors

RICHARD HEINBERG is Senior Fellow-in-Residence of Post Carbon Institute and is widely regarded as one of the world's foremost educators on the need to transition away from fossil fuels. He is the author of twelve books, including seminal works on society's sustainability crisis, *The Party's Over: Oil, War & the Fate of Industrial Societies* (2003) and *The End of Growth: Adapting to Our New Economic Reality* (2011). He has authored scores of essays and articles that have appeared in *Nature, Christian Science Monitor*, the *Wall Street Journal*, and elsewhere; has been quoted and interviewed countless times for print, television, and radio; and has spoken to hundreds of audiences in fourteen countries.

David Fridley is a staff scientist at Lawrence Berkeley National Laboratory (LBNL), where he is deputy group leader of the China Energy Group. His work has involved extensive collaboration with China on end-use energy efficiency and modeling, industrial energy use, energy policy research, low-carbon city development, and energy supply assessment. He has published dozens of articles in peer-reviewed journals and authored chapters in three books. Prior to joining LBNL he was a consultant on downstream oil markets in the Asia-Pacific region and a business development manager for Caltex China. He is a Fellow of Post Carbon Institute.

About Post Carbon Institute

Post Carbon Institute envisions a transition to a more resilient, equitable, and sustainable world. It provides individuals and communities with the resources needed to understand and respond to the interrelated ecological, economic, energy, and equity crises of the twenty-first century. Visit postcarbon.org for a full list of fellows, publications, and other educational products.

For supplementary content and resources for this book, visit **OurRenewableFuture.org**.